T0174411

Active and Programmable Networks for Adaptive Architectures and Services

OTHER TELECOMMUNICATIONS BOOKS FROM AUERBACH

Architecting the Telecommunication Evolution: Toward Converged Network Services
Vijay K. Gurbani and Xian-He Sun
ISBN: 0-8493-9567-4

Business Strategies for the Next-Generation Network
Nigel Seel
ISBN: 0-8493-8035-9

Chaos Applications in Telecommunications
Peter Stavroulakis
ISBN: 0-8493-3832-8

Context-Aware Pervasive Systems: Architectures for a New Breed of Applications
Seng Loke
ISBN: 0-8493-7255-0

Fundamentals of DSL Technology
Philip Golden, Herve Dedieu, Krista S Jacobsen
ISBN: 0-8493-1913-7

Introduction to Mobile Communications: Technology, Services, Markets
Tony Wakefield
ISBN: 1-4200-4653-5

IP Multimedia Subsystem: Service Infrastructure to Converge NGN, 3G and the Internet
Rebecca Copeland
ISBN: 0-8493-9250-0

MPLS for Metropolitan Area Networks
Nam-Kee Tan
ISBN: 0-8493-2212-X

Performance Modeling and Analysis of Bluetooth Networks: Polling, Scheduling, and Traffic Control
Jelena Misic and Vojislav B Misic
ISBN: 0-8493-3157-9

A Practical Guide to Content Delivery Networks
Gilbert Held
ISBN: 0-8493-3649-X

Resource, Mobility, and Security Management in Wireless Networks and Mobile Communications
Yan Zhang, Honglin Hu, and Masayuki Fujise
ISBN: 0-8493-8036-7

Security in Distributed, Grid, Mobile, and Pervasive Computing
Yang Xiao
ISBN: 0-8493-7921-0

TCP Performance over UMTS-HSDPA Systems
Mohamad Assaad and Djamal Zeghlache
ISBN: 0-8493-6838-3

Testing Integrated QoS of VoIP: Packets to Perceptual Voice Quality
Vlatko Lipovac
ISBN: 0-8493-3521-3

The Handbook of Mobile Middleware
Paolo Bellavista and Antonio Corradi
ISBN: 0-8493-3833-6

Traffic Management in IP-Based Communications
Trinh Anh Tuan
ISBN: 0-8493-9577-1

Understanding Broadband over Power Line
Gilbert Held
ISBN: 0-8493-9846-0

Understanding IPTV
Gilbert Held
ISBN: 0-8493-7415-4

WiMAX: A Wireless Technology Revolution
G.S.V. Radha Krishna Rao, G. Radhamani
ISBN: 0-8493-7059-0

WiMAX: Taking Wireless to the MAX
Deepak Pareek
ISBN: 0-8493-7186-4

Wireless Mesh Networking: Architectures, Protocols and Standards
Yan Zhang, Jijun Luo and Honglin Hu
ISBN: 0-8493-7399-9

Wireless Mesh Networks
Gilbert Held
ISBN: 0-8493-2960-4

AUERBACH PUBLICATIONS
www.auerbach-publications.com
To Order Call: 1-800-272-7737 • Fax: 1-800-374-3401
E-mail: orders@crcpress.com

Active and Programmable Networks for Adaptive Architectures and Services

Syed Asad Hussain

Auerbach Publications
Taylor & Francis Group
Boca Raton   New York

Auerbach Publications is an imprint of the
Taylor & Francis Group, an informa business

Auerbach Publications
Taylor & Francis Group
6000 Broken Sound Parkway NW, Suite 300
Boca Raton, FL 33487-2742

© 2007 by Taylor & Francis Group, LLC
Auerbach is an imprint of Taylor & Francis Group, an Informa business

No claim to original U.S. Government works
Printed in the United States of America on acid-free paper
10 9 8 7 6 5 4 3 2 1

International Standard Book Number-10: 0-8493-8214-9 (Hardcover)
International Standard Book Number-13: 978-0-8493-8214-7 (Hardcover)

Library of Congress Cataloging-in-Publication Data

Hussain, Syed Asad.
    Active and programmable networks for adaptive architectures and services /
Syed Asad Hussain.
        p. cm.
    ISBN 0-8493-8214-9 (alk. paper)
    1. Computer networks--Management. I. Title.

TK5105.5.H876 2006
004.6--dc22   2006047731

Visit the Taylor & Francis Web site at
http://www.taylorandfrancis.com

and the Auerbach Web site at
http://www.auerbach-publications.com

Dedication

Dedicated to those who are firmly rooted in knowledge.

No exaltation or grandeur is superior to learning and knowledge.

Hazrat Ali (A.S.)

Contents

Preface

New applications such as video conferencing, video on demand, multimedia transcoders, Voice-over-IP (VoIP), intrusion detection, distributed collaboration, and intranet security require advanced functionality from networks beyond simple forwarding congestion control techniques. Examples of advanced functionality include self-reconfiguration, traffic monitoring and analysis, distributed and secure communication, and the ability to adjust to application requirements through deployment of new services. Traditional network devices such as routers and switches are closed, vertically integrated systems. Their functions are rigidly programmed into the embedded software and hardware by the vendors. Their functions are usually limited to simple management, routing, congestion control, etc. The traditional architectures often have difficulty integrating new technologies and standards into the shared network infrastructure. The new services can dynamically extend the capabilities of the existing networking architectures.

Active and programmable networks allow the creation, customization, deployment, and management of new services or applications that are deployed (programmed) dynamically into network nodes. Users are thus able to utilize these programmable services to attain their required network support in terms of performance and flexibility.

This book clearly and comprehensively explains the concept of active and programmable networks. It deals with the current areas of research in active and programmable networks. The research areas include active packet scheduling, routing, network management, wireless networks, and security. It also provides a deeper insight into the architectures and working of active and programmable networks for students and researchers who seek challenging tasks that extend frontiers of technology. At the end, it has a complete section on modeling and simulation of active and programmable networks.

This book should be of considerable use for communications and networking engineers, teachers and students, and particularly for forward-looking companies that wish to actively participate in the development of active networks and desire to ensure a head start in the integration of this technology in their products.

Chapter 2 describes the general architecture for active and programmable networks. It also presents quality of service (QoS) technologies for Internet Protocol (IP) networks and the Institute of Electrical and Electronics Engineers (IEEE) 1520 standard for programmable networks.

Chapter 3 elaborates on enabling technologies for programmable networks. It discusses in detail agents, middleware issues, dynamically reconfigurable hardware, and operating systems. Chapter 4 presents a detailed description of certain active and programmable paradigms. Chapter 5 is based on scheduling schemes. Chapter 6 deals with management architectures for active and programmable networks. It also discusses Simple Network Management Protocol (SNMP). Chapter 7 describes programmable routing schemes. It discusses in detail different active multicasting mechanisms, such as active gathercast and active reliable multicast. There is a section on active and programmable router architectures as well. Chapter 8 presents different active wireless and mobile solutions for traditional wireless and mobile networks. It discusses the concept of active base stations and programmable handoffs. It also consists of a section on adaptive management architecture for ad hoc networks. Chapter 9 deals with the security issues in active and programmable networks, and Chapter 10 describes certain areas where the concepts of active and programmable networks have been applied.

I express my gratitude to my wife, daughter, and family members for their patience and encouragement during the preparation of this book. I am grateful to my students Khawar Mehmood and Abdul Basit for their help in the preparation of some chapters. I am thankful to Auerbach Publications (Taylor & Francis Group) for providing me an opportunity to write this book. Finally, I thank Mr. Richard O'Hanley for providing me with the necessary guidelines regarding the preparation of this book.

Syed Asad Hussain

About the Author

Syed Asad Hussain obtained his Ph.D. from Queen's University, Belfast, U.K., and his M.Sc. from the University of Wales, Cardiff, U.K. Presently, Dr. Hussain is an assistant professor in the Department of Computer Science at COMSATS Institute of Information Technology, where he is leading the research on networks. Previously, he worked as an engineer at Paktel, a cable and wireless company.

His interests are in the areas of active and programmable networks, wireless and mobile networks, and network modeling and simulation. He has published several research papers in the areas of computer networks and telecommunications. A member of IEEE, he has served on technical program committees and on organizing committees of several conferences. He also regularly reviews papers of several international journals.

Chapter 1

Introduction

There have been several advancements in communication systems in general and telecommunication systems in particular in the last decade. The speed and capacity of various components in a telecommunication system, such as transmission media, switches, memory, and processors, have increased exponentially. The advent of high-speed networking has introduced opportunities for efficient transfer of applications such as videoconferencing, video on demand, and Voice-over-IP along with data applications. These applications have stringent performance requirements in terms of throughput, end-to-end delay, delay jitter, and loss rate.

Traditionally, networks have been used to deliver packets from one endpoint user to another. In this case, there has been a distinct boundary between the functions inside a network and what users do. The user data is transferred passively from one end to another. The network is insensitive to the user bits, and they are transferred without modifications. The role of network elements as far as the computation is concerned is limited. Today's networks are the result of decades of innovative thinking and engineering, and these are functioning admirably well. Examples of this success are the telephone and Internet. If these networks have worked successfully for a long time, then why adopt a drastically different approach?

The telephone was invented more than 100 years ago, and most people use this basic service (with some additional services). The Ethernet protocol was developed some 25 years ago. The Transmission Control Protocol/Internet Protocol (TCP/IP) suite was also designed 20 years ago. The continuous use of these network technologies and protocols is a testament to their original design, but on the other hand, it shows that the networks have evolved slowly. This is due to the reasons of interoperability, i.e., protocols must be agreed upon through standardization. The network providers must then wait for vender implementations and then deploy new equipment in their networks. Lastly, subscribers see new services offered. In the past, while the network evolution was slow, people were satisfied with the basic voice and data services, and the telecommunications infrastructure was not complex and sophisticated. The explosive growth and commercialization of the Internet have created demands for new services and application. In this situation, service providers have to respond more quickly and dynamically than they have traditionally. The service and network providers cannot wait for gradual vendor implementations.

As computing power becomes cheaper, more and more functionality is deployed into network processing elements. Examples of such functionality are congestion control, packet filtering, etc.

In the present-day Internet, the intermediate nodes (e.g., routers and switches) are closed systems whose functions are rigidly programmed into the embedded software and hardware by the vendors. The drawbacks of this approach are a long standardization process for the development and deployment of new technologies and protocols into the shared network infrastructure, poor performance due to redundant operations at several protocol layers, and difficulty accommodating new services in the existing architectural model. Thus, the introduction of new services is a challenging task, requiring new tools for service creation, including new network programming platforms and supporting technologies.

An approach known as *active and programmable networks* has emerged to address these issues. Active and programmable networks allow dynamic customization of nodes, thus allowing the creation of new network architectures.[1] The key aim of active and programmable networks is to enable the addition of user or agent code into network elements to be a part of the normal operation of the network, thus allowing new functionality to be rapidly introduced into the network, perhaps on the timescale of a single session or even a packet. Active and programmable networks seek to exploit advanced software techniques and technologies, e.g., software agents and middleware such as Common Object Request Broker Architecture (CORBA), to make network infrastructures more flexible, thereby allowing end users, network operators, or service providers to customize network elements to meet their specific needs.[2] Thus, future open or programmable networks are likely to be based on active networking agent technologies and open signaling techniques.[2] The aim of these techniques is to open up the network and accelerate its programmability in a controlled manner for the deployment of new architectures and services.

1.1 A Brief Networking History

The major factor in the evolution of the computer networking industry is the growth of the Internet. Today's Internet can be traced back to the ARPANet,[3] developed in 1969 under a contract allowed by the Advanced Research Projects Agency (ARPA), which initially connected four major computers at universities in the southwestern United States (UCLA, Stanford Research Institute, UCSB, and the University of Utah). Although networking research in Europe first started in the late 1970s, it was mainly confined to developments of national research networks. The contract was carried out by BBN of Cambridge, MA, under Bob Kahn and went online in December 1969. By June 1970, MIT, Harvard, BBN, and Systems Development Corp. (SDC) in Santa Monica, CA, were added. By January 1971, Stanford, MIT's Lincoln Labs, Carnegie Mellon, and Case-Western Reserve University were added. Later on, NASA/Ames, Mitre, Burroughs, RAND, and the University of Illinois joined in. After that, the listing kept on increasing. The ARPANet was designed in part to provide a communications network that would work even if some of the sites were destroyed by nuclear attack. If the most direct route was not available, traffic would be directed around the network via alternate routes.

E-mail was adapted for ARPANet by Ray Tomlinson of BBN in 1972. He picked the @ symbol from the available symbols on his teletype to link the username and address. The Telnet protocol, enabling logging on to a remote computer, was published as a Request for Comments (RFC) in 1972. RFCs are a means of sharing developmental work throughout the community. The File Transfer Protocol (FTP), enabling file transfers between Internet sites, was published as an RFC in 1973, and from then on RFCs were available electronically to anyone who had use of the FTP. The Internet matured in the 1970s as a result of the TCP/IP architecture first proposed by Bob Kahn at BBN and further developed by Kahn and Vint Cerf at Stanford and others throughout the 1970s. It was adopted by the Defense Department in 1980, replacing the earlier Network Control Protocol (NCP), and was universally adopted by 1983.[4]

The UNIX to UNIX Copy Protocol (UUCP) was invented in 1978 at Bell Labs. Usenet was started in 1979 based on UUCP.[4] Newsgroups, which are discussion groups focusing on a topic, followed, providing a means of exchanging information throughout the world. Although Usenet is not considered part of the Internet, because it does not share the use of TCP/IP, it linked UNIX systems around the world, and many Internet sites took advantage of the availability of newsgroups. It was a significant part of the community building that took place on the networks.

In 1986, the National Science Foundation funded NSFNet (National Science Foundation Network) as a cross-country 56-Kbps backbone for the Internet. It maintained its sponsorship for nearly a decade, setting rules for NSFNet's noncommercial government and research uses.

As the commands for e-mail, FTP, and Telnet were standardized, it became a lot easier for nontechnical people to learn to use the networks. It was not easy by today's standards, but it did open up use of the Internet to many more people, in universities in particular. Other departments besides the libraries, computer, physics, and engineering departments found ways to make good use of the networks to communicate with colleagues around the world and to share files and resources.

In 1989 another significant event took place in making networks easier to use. Tim Berners-Lee and others at the European Laboratory for Particle Physics, more popularly known as CERN, proposed a new protocol for information distribution. This protocol, which became the World Wide Web in 1991, was based on a hypertext system of embedding links in text to links to other text.

The development in 1993 of the graphical browser Mosaic by Marc Andreessen and his team at the National Center for Supercomputing Applications (NCSA) gave the protocol its big boost. Later, Andreessen moved to become the brain behind Netscape Corp., which produced the most successful graphical browser and server until Microsoft launched Microsoft Internet Explorer.

Because the Internet was initially funded by the government, it was originally limited to research, education, and government uses. Commercial uses were prohibited unless they directly served the goals of research and education. This policy continued until the early 1990s, when independent commercial networks began to grow. It then became possible to route traffic across the country from one commercial site to another without passing through the government-funded NSFNet Internet backbone.

Delphi was the first national commercial online service to offer Internet access to its subscribers.[4] It opened up an e-mail connection in July 1992 and full Internet service in November 1992. All limitations on commercial use disappeared in May 1995 when the National Science Foundation ended its sponsorship of the Internet backbone, and all traffic relied on commercial networks. AOL, Prodigy, and CompuServe came online. Because commercial usage was so widespread by this time and educational institutions had been paying their own way for some time, the loss of NSF funding had no appreciable effect on costs.[4]

Today, NSF funding has moved beyond supporting the backbone and higher educational institutions to building the K–12 and local public library accesses on the one hand, and the research on the massively high volume connections on the other.

Microsoft's full-scale entry into the browser, server, and Internet service provider market completed the major shift over to a commercially based

Internet.[4] The release of Windows 98 in June 1998 integrated well into the market. Later, Microsoft launched Windows 2000 and Windows XP.

A current trend with major implications for the future is the growth of high-speed connections. 56K modems were not fast enough to carry multimedia, such as sound and video, except in low quality. But new technologies many times faster than 56K modems, such as cable modems, Digital Subscriber Lines (DSLs), and satellite broadcast, are available now.

During this period of enormous growth, businesses entering the Internet arena scrambled to find economic models that worked.[4] Free services supported by advertising shifted some of the direct costs away from consumers temporarily. Services such as Delphi offered free Web pages, chat rooms, and message boards for community building. Online sales have grown rapidly for such products as books and music CDs and computers, but the profit margins are slim when price comparisons are so easy, and public trust in online security is still shaky. Business models that have worked well are portal sites, which try to provide everything for everybody, and live auctions. AOL's acquisition of Time-Warner was the largest merger in history when it took place and shows the enormous growth of Internet business.[4] The stock market has had a rocky ride, swooping up and down as the new technology companies, the dot coms, encountered good news and then bad news. The decline in advertising income spelled doom for many dot coms.[4]

A major pan-European cooperation in the networks started with the establishment of the RARE (Réseaux Associés pour la Recherche Européenne/ European Association of Research Networks) organization in 1986.[5] The first real attempt to define a longer-term set of objectives and goals for European research networking was the COSINE (Co-operation for Open Systems Interconnection in Europe) project.[5] COSINE had the aims of improving cooperation among research networks in Europe while at the same time promoting the development of Open System Interconnect (OSI). It therefore had too many different targets to represent a strategic direction for European research networking. A more focused approach was required.

The national research networking organizations, although grouped together within RARE, still needed an efficient and cost-effective vehicle to coordinate pan-European research networking on their behalf, and to ensure that project results were delivered on time, within the budget, and with high levels of reliability.

After two years of preparations, DANTE was launched on July 6, 1993, at St. John's College in Cambridge in the U.K.[6] Its aim was to organize the management of otherwise fragmented, uncoordinated, expensive, and inefficient transnational services and operational facilities.

During the first year of DANTE's existence, RARE was the legal owner and only shareholder. Then on March 25, 1994, the ownership of the company was formally transferred to 11 national research networking organizations. There have been some small changes and four additions to the shareholders' list over the years.

Following from the International X.25 Interchange (IXI) initiative, which was part of the COSINE project, DANTE managed the EuropaNET project. EuropaNET was the first generation of pan-European research networks to be managed by DANTE, and the company has gone from strength to strength since then.

Since its creation in 1993, DANTE has played a pivotal role in the formation and management of four consecutive generations of the pan-European research network: EuropaNET, TEN-34, TEN-155, and GÉANT. All these networks have been established and supported in the context of European Union programs, such as the Fourth and Fifth Framework Programmes and eEurope. In addition, DANTE has managed or been a partner in numerous other research networking projects.

From 1993 to 1997, EuropaNET was developed. It connected 18 countries at speeds of 2 Mbps and used IP technology.[6]

- From 1997 to 1998, TEN-34 was developed. It connected 18 countries. The speed was 34 Mbps and it used both IP and Asynchronous Transfer Mode (ATM) technology.
- From 1998 to 2001, TEN-155 was developed, connecting 19 countries at speeds of 155 to 622 Mbps and again using IP and ATM technologies.
- From 2001 until 2004, the GÉANT network connected 32 countries at speeds of 2.5 to 10 Gbps. It used dense wavelength division multiplexing (DWDM) technology and offered both IPv4 and IPv6 native services in dual-stack mode.

The development of each generation of network has typically been undertaken as a project involving a consortium of National Research and Education Networks (NRENs), with DANTE acting as a managing or coordinating partner.

In addition to improving pan-European research network connectivity, these networks have been used to conduct a number of test programs, focusing primarily on ATM and quality of service (QoS). These have been carried out by task forces, such as Task Force TEN (TF-TEN), TF-TANT (Testing of Advanced Networking Technologies), and TF-NGN (New Generation Networks).

1.2 Network Standards and Protocols

Standards and protocols allow computers and devices from different vendors to connect and communicate with each other. Standardized documents speed up the transfer of goods or identify sensitive or dangerous goods that may be handled by people speaking different languages. Standardization of connections and interfaces of all types facilitates the compatibility of equipment of diverse origins and the interoperability of different technologies.[7–12]

Standards and protocols are two widely used and frequently confused terms in the field of computer networks. A protocol is a set of rules and formats that govern the communication between peer entities within a layer.[12] An entity is any object capable of sending or receiving information. Protocols allow peer entities to agree on a set of operations and the meaning of each operation. An important function of a protocol is to describe the *semantics* of a message, that is, the meaning of each section of bits.

Protocols can be alternatively described as providing a *service*, such as reliable file transfer or e-mail transfer. In other words, a protocol is used by a peer entity to provide a service to a higher-level layer entity. Hence, service is a set of operations performed between layers vertically. The service identifies what operations a layer performs on behalf of its users, but hides the implementation details of these operations. In a *layered* architecture, lower layers provide service to upper layers, as shown in Figure 1.1.

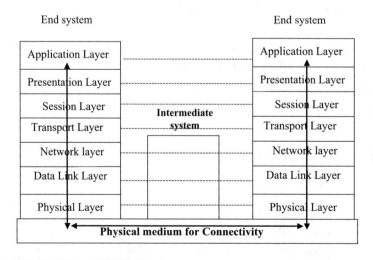

Figure 1.1   Layered architecture.

The layers basically show a dependency on each other for the implementation of services through protocols, for example, transfer of e-mail messages through the Simple Mail Transfer Protocol (SMTP). In SMTP, the sender establishes a TCP connection to port 25 of the destination machine. After establishment of the TCP connection to port 25, the sender (client) waits for the receiving machine (server) to respond. Next, the server identifies itself and tells whether it is ready to receive mail. If it is prepared to receive mail, the sender sends the message and the server acknowledges it. If the server is not prepared to accept the mail, the sender releases the connection. The TCP layer provides a reliable connection for the transfer of data between the client and the server. The Internet layer offers the routing function across multiple networks. This example shows that e-mail message transfer involves multiple layers, and these layers are dependent on each other for complete transfer. Packets exchanged between peer entities at the same layer are called protocol data units (PDUs).[7,8] The packets transferred to a layer by an upper layer are called service data units (SDUs). A PDU of a layer contains a header and possibly user data of that layer. The header in each PDU consists of control information by peer entities at a layer. The header includes the following information:[7–9]

> **Destination port**: To which client the packet is to be delivered.
> **Sequence number**: If packets arrive out of order, sequence numbers are used by destination entity to reorder them.
> **Error detection**: Error detection calculation is performed at the destination, and results are compared with the incoming data. A difference in results shows that there has been an error.

1.3 Protocol Reference Models

This section discusses the two well-known network reference models: the OSI model[10–15] and the TCP/IP reference architecture.[3,16,17]

1.3.1 The OSI Model

The Open System Interconnect (OSI) was developed by the International Standards Organization (ISO). This model is called Open System Interconnect because it deals with connecting open systems. The protocols' details are open to the public, as well as changes to the protocols. The task of ISO was to define a set of layers and the services provided by each layer. Each layer performs a particular set of functions. The idea behind the logical function grouping into layers was to divide a complex communication problem into smaller, manageable tasks. The other

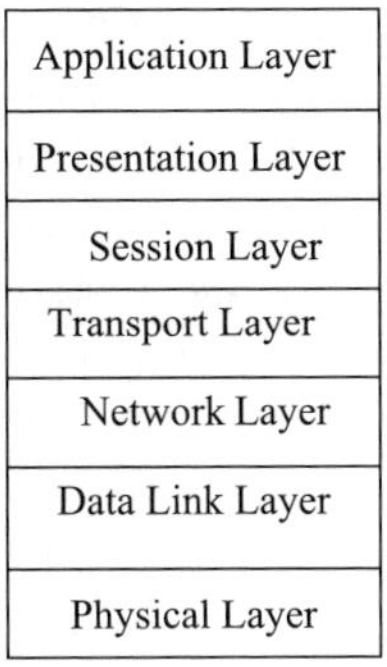

| Application Layer |
| Presentation Layer |
| Session Layer |
| Transport Layer |
| Network Layer |
| Data Link Layer |
| Physical Layer |

Figure 1.2   The OSI seven-layer model.

consideration was to keep the number of layers to a reasonable number so that the processing overhead imposed by the collection of layers is not burdensome. The OSI model consists of seven layers, as shown in Figure 1.2.

The two end systems consist of all seven layers, but intermediate systems such as switches and routers implement only the lowest three layers.[12] The principles used in designing seven layers can be briefly summarized as follows:

- The number of layers is kept low so that their description and management remain easy.
- Each layer carries out a well-defined function, and the number of interactions across the boundary are minimal.
- Changes of functions or protocols are allowed within a layer without affecting other layers.
- Similar functions are collected into the same layer.
- International standardization of protocols and interfaces is kept in mind while creating a boundary or defining functions of a layer.

The seven layers of the OSI model are discussed below. The description starts from the bottom layer.

1.3.1.1   Physical Layer

The physical layer is concerned with the transmission of raw bit stream over physical media.[11,19] It deals with the mechanical and electrical specifications of interface and transmission media. In context of the Internet, the physical layer provides the media for transporting these bits. The transmission media include coaxial cables, twisted-pair cables, satellite transponders and earth stations, and optical fiber links with associated optical transmitters and receivers. The physical layer provides services to the data link layer. The data in the data link layer consists of data in the form of 1s and 0s organized in frames to be sent across the transmission medium. The physical layer converts this stream of 1s and 0s into signals that are transported across the transmission medium. In addition to these services, the physical layer provides bit rate control, bit synchronization, multiplexing, etc.

1.3.1.2 Data Link Layer

The data link layer organizes the data received from the physical layer in the form of frames.[10–12] It is responsible for carrying a frame from one node to the next (on the hop). This is in contrast to the network layer, which routes a packet through a series of nodes (called routers). It provides not only framing, but also the functionality of error and flow control. That is, it makes sure that a packet is received without any error. If the packet is corrupted on the link, it should be either corrected or retransmitted. The data link layer must also take care of flow control, that is, control the rate at which packets are placed on a hop. The other two issues addressed by the data link layer are medium access control and addressing.

The addresses in a broadcast system, like local area network (LAN) environments, make sure that the node should receive the packet addressed to it. The data link layer ensures this functionality by allocating unique data link layer addresses to nodes. These addresses are called physical addresses or *Medium Access Control* (MAC) addresses. Because multiple nodes share a common medium (guided or wireless), we need a way to control access to the medium at any time. A sublayer of the data link layer, called the MAC layer, controls access to the medium by allowing the nodes to transmit in time slots (e.g., Distributed Interframe Space (DIFS)). The flow control and error control functions are provided by another sublayer of the data link, called the *logical link control*.

1.3.1.3 Network Layer

The network layer transfers a packet from one terminal to another; it is responsible for host-to-host delivery.[12] The network layer lies between the data link layer and transport layer. It receives data from the data link layer and delivers it to the transport layer. The data link is responsible for the transfer of data from node to node. The data may travel several nodes or *subnets* before reaching the destination. The network layer makes sure that these packets are transferred to the destination through several subnets

(Internet). The main functions of the network layer are internetworking, addressing, routing, packetizing, and fragmentation.

Internetworking is the logical attachment of different heterogeneous subnets together to give a feeling of a single network to the user applications. The devices on the Internet can communicate globally only if they are uniquely identified by an address. Two devices cannot have the same address. Because the devices are connected to the Internet, these addresses are called *Internet addresses* or *IP addresses*. A network is divided into several smaller networks called *subnets*. This process is called subnetting. The IP address is divided into different levels of hierarchy. The first part is the network identifier (NetID) of the network, and the second part shows the host on the network.

The *datagrams* are packets routed by the network layer across the Internet. There are many routing protocols. The Internet is a combination of many networks, and any routing protocol can be selected depending on the type of efficiency required from the network. The network layer encapsulates the packets received from the transport layer and creates new packets for onward transfer. This process is packetization. Depending on the type of physical networks, the datagrams are divided into smaller units. Each unit has its own header. This process is called fragmentation.

1.3.1.4 Transport Layer

The transport layer lies between the network layer and the session layer. The transport layer provides reliable, flow-controlled, error-free end-to-end service.[12] Error control techniques are used to recover packet loss, damage, and duplication problems. The lost packets are retransmitted; damaged packets are detected, discarded, and retransmitted. The duplication problem is solved by detecting and discarding the duplicated packets.

The other important function of transport layer is flow control. The flow control at this layer is performed end-to-end, unlike in the data link layer, which provides flow control on a per hop basis. The transport layer provides two types of services to the session layer. The Transmission Control Protocol (TCP) provides connection-oriented service, and the User Datagram Protocol (UDP) provides connectionless service to its applications. TCP ensures guaranteed delivery of applications to the destination. TCP provides error detection and correction in addition to flow control and multiplexing to applications like the World Wide Web, file transfer, and e-mail. The TCP transmits data in the form of segments, and each segment consists of packets. The packets in UDP are not numbered; they may be delayed, lost, or arrive out of sequence. They are not acknowledged either. Then why use UDP? Because UDP is a simple protocol with

Source Port	Destination port
Sequence Number	
Acknowledgment Number	

Header Length	Unused	Flags	Window

Checksum	Urgent Pointer
Options + Padding	

Source Port	Destination Port
Segment Length	Checksum

Figure 1.3   The headers of TCP and UDP.

a minimum of overhead. Figure 1.3 shows the headers of TCP and UDP. UDP is faster than TCP because it requires less interaction between a sender and receiver. Real-time applications like voice and video applications use UDP. These services can sustain packet loss to some extent, but they are sensitive to delays.

1.3.1.5   Session Layer

The session layer is responsible for the establishment, management, and termination of sessions between two machines.[10,12] It also provides synchronization functionality, that is, it allows applications to start from where they were after a crash.

1.3.1.6   Presentation Layer

The presentation layer is concerned with the representation of data transmitted.[12] It makes it possible for clients with different data types to communicate, that is, it hides the differences between applications. The presentation layer also performs encryption, authentication, and data compression. Encryption and authentication are security features of networks.

1.3.1.7 Application Layer

The application layer is responsible for providing distributed application services.[11,13] It consists of many protocols, for example, Hypertext Transfer Protocol (HTTP) to support World Wide Web (WWW), SMTP to transfer e-mail, and FTP to transfer files. The application layer utilizes the services of the six layers above, but it does not provide services to any other layer.

1.3.2 Why Are Protocol Reference Models Layered?

The layering of protocol reference models provides a structure for standardization. The protocol layering smoothes the process of standardization in two ways:[19] The first advantage is that it provides us with services by dividing a complex task into simpler ones.[19] For example, consider electronic mail service. A host runs a user agent to compose and read messages. The user, after writing an e-mail message and providing a destination address to the message, just clicks on an icon to deliver to the destination. The user agent transfers the message to the message transfer agent on the user's host. The message transfer agent facilitates its delivery by using Domain Name System (DNS) to find the mail server willing to accept the message. The transfer agent now finds the IP address of this mail server using DNS. It then establishes a TCP connection to the SMTP server on port 25 of the mail server. The message is transferred to the recipient's mailbox by using the Simple Mail Transfer Protocol (SMTP); the message transfer agent then breaks the TCP connection. Layering in this case decomposes the complicated task into simpler ones. The functions of connection establishment and message transfer in this example are transparent to the user. The user simply types the e-mail message and presses the Send icon. Each layer performs its own task.

The second advantage is that the implementation details of layers are hidden from other layers.[19] Because the implementation details of layers are hidden from each other and the boundaries between layers are well defined, we can make changes in the layers without affecting other layers in the protocol stack. For example, the packet-switching technique has migrated from X.25 to frame relay and from frame relay to ATM. The replacement of these technologies with the combination of ATM and the synchronous optical network (SONET) at the lower layers (physical and data link) to improve speed allows high-speed data transfer between users without affecting the upper layers (application layer, etc.) of the protocol stack.

1.3.3 Drawbacks of the OSI Model

It is considered that the Internet protocol or TCP/IP model has won over the OSI model in implementation and usage. Following are the main drawbacks of the OSI model:

1. The OSI standardization process was slow and complex. For example, multicast service was initiated through the Internet very quickly. The OSI model provides provisions for point-to-point communication only.

2. The OSI layers lead to information hiding. Though information hiding is useful, it can sometimes lead to poor performance.[20] For example, the transport layer slows down the source after detecting a packet loss. The transport layer does not know what is going on at the network layer and how packets are transferred across the network. Suppose the packet loss is due to link errors rather than congestion on the network. In this situation, the transport layer wrongly throttles the source. If the network layer had informed the transport layer, the transport layer could have differentiated between the packet losses due to congestion and link errors. This situation could have avoided the unnecessary throttling of the source. Good designs can avoid such types of problems.

3. The existence of entities in the next higher layer is assumed at connection setup time. This means that before a CONNECT indication can occur in the transport layer, creation of an entity in the session layer has to be performed.[3]

4. The quality of service (QoS) is negotiated only at the connection setup phase. With the increase in the demand of multimedia services on the Internet, there are instances where dynamic change of the QoS is required. In the OSI model, QoS can only be changed by terminating the existing connection and then reestablishing a new connection with the new QoS. This policy is too rigid, inefficient, and costly in terms of resource consumption. The OSI model does not even specify the criteria of mapping different QoS at different layers. For example, it does not tells us how the retransmissions at the transport layer affect the delays at the network layer.

5. Some of the QoS specified in the ISO draft proposals DP8348 and DP8073 as parameters of the primitives are difficult to understand for a user.[14] For example, the meaning of the negotiation of the connection establishment failure probability at the connection setup time or of the DISCONNECT failure probability is not very clear for the user of the service.[3]

6. Remote procedure call (RPC) is an important mechanism to achieve transparency in distributed network communication. This model hides network communication between machines by allowing a process to call a procedure on a remote machine, e.g., a server. When a procedure is called, the parameters are transported transparently to the remote machine (e.g., a server) where the procedure is executed and results are communicated back to the caller. It appears to the client that the procedure call is executed locally. In the OSI and TCP/IP models, RPC is treated as a high-level service because of its needs for support by other services (data integrity, addressing, and presentation).[22] These models therefore place RPC on the top of their protocol stacks. Distributed operating systems[21] consider RPC to be a single optimized and monolithic protocol that incorporates all services (data integrity, addressing, presentation, etc.), thus violating the concept of separate layers, as mentioned in these two models.

7. Data manipulation, e.g., encryption, presentation, formatting, compression, and computation of checksum, is computationally complex and processing intensive. This is due to the reason that reading and writing each byte of data in a message consists of memory loads and stores. Clark and Tennenhouse[9,23] propose integrated layer processing (ILP) as a solution to this problem. The layer integration (presentation and application layers) can improve the throughput by reducing the number of loads and stores. What happens is that a message is passed from one protocol to the next as a complete message in the form of a data structure. Any protocol that operates on the data in the messages must load and possibly store each byte of the message. Integrated layer processing combines data manipulation operations from a series of protocols into a *pipeline* that provides shared accesses to the message data structure. The relationship between resequencing and presentation processing is in fact one of the key architectural considerations in layered protocol design. In the OSI model and also in the TCP/IP model, the presentation conversion cannot be performed in real-time with data resequencing and recovery. The data gets out of order because packets are reordered in some intermediate switching nodes. The other reason is that if a packet is lost and retransmitted and in the mean time other packets have arrived, it will be considered out of order. Applications cannot deal with packet loss or reordering. Instead, lower layer protocols such as TCP buffer all the follow-up data, request retransmission of the lost packet, and then proceed with final manipulation when it arrives. In this situation, any presentation conversion can only occur after TCP

has completely reordered and recovered the incoming packets. Hence, reordering of packets temporarily stops presentation conversion. Many data manipulations can only be performed once the data unit is in order. This is true of most error detection checksums, encryption schemes, most presentation transformations, and when moving the data to or from the user address space. Thus, it is ensured by the protocol that the data is in order, at least within a certain range, before performing these manipulations.

1.3.4 Ordering Constraints

Layered protocol suites provide isolation between the functional units of distinct layers. A major architectural advantage of isolation is that it facilitates the implementation of subsystems whose operation is restricted to a small subset of protocol stacks. The implementation of a layered protocol stack involves the *sequential* processing of each unit of information as it is passed down the individual protocol layer entities and passed up through the peer entities of the protocol stack. This sequential approach induces the problem of *ordering constraints*.[9] That is, the protocols should be implemented in the order they appear on the protocol stack. Tasks within a protocol are subject to internal ordering constraints. For example, a checksum protocol cannot write the checksum result into the message header (header processing) until after it has computed a checksum on the data. Another example is of ordering constraints between layers. For example, an encryption protocol at the presentation layer must decrypt higher-level headers (header processing) in a received message before the next layer can read its header (header processing). These ordering constraints are implemented by the serial execution of layers, but not when layers are integrated.[23] Message rejection is a potential problem for protocol integration, because a message may be in the middle of an integrated series of protocols when it is rejected. Layers that are logically subsequent to the rejecting layer may have already begun processing the message. Additionally, layers that logically precede the rejecting layer may not have completed processing the message. The solution was presented by Abbott and Peterson.[23] They executed the send and deliver operations in three stages: initial stage (header processing for delivery), data manipulation stage (transfer of data, error detection, buffering for retransmission, encryption, presentation formatting, moving data to or from address space), and final stage (header processing for sending). The initial and final stages are executed serially and the data manipulations take place in an integrated fashion in one shared stage, as shown in Figure 1.4.

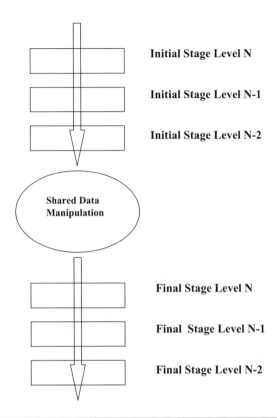

Figure 1.4   Integrated data manipulations.

Message processing tasks are performed in the appropriate stages to satisfy the ordering constraints. Within a stage, a protocol is free to perform the tasks of that stage in any order. Message rejection is deferred until the final stage.

1.3.5 Other Factors

The operating system adds overhead to packet processing. Packet processing requires considerable support from the operating system. The necessary actions are to take an interrupt, allocate a packet buffer, free a packet buffer, restart the input/output (I/O) device, wake up a process, and restart a timer.

In TCP, packets are coupled to timers. A retransmit timer is set on after sending the data. On receipt of an acknowledgment, this timer is cleared. The overhead of managing these timers also constitutes an overhead to packet processing.

Table 1.1   The Measured Overheads of TCP

Per Byte	
User system copy	200 μs
TCP checksum	185 μs
Network memory copy	386 μs
Per Packet	
Ethernet driver	100 μs
TCP + IP + ARP	100 μs
Operating system overhead	240 μs

Source: Clark, D.D. et al., *IEEE Commun. Mag.*, Vol. 27, pp. 23–29, 1989. (© 1989, IEEE)

Another type of overhead is associated with the operations that touch the data bytes. The example is checksum computation. The movement of data in memory also adds overhead to the protocol processing.

Data is moved in memory for two reasons. First, it is moved to separate the data from the header and get the data into the alignment needed by the application. Second, it is copied to get it from the I/O device to system address space and user address space. D.D. Clark et al. performed different tests to calculate TCP processing overhead. Table 1.1 shows the measured overheads.[10]

1.3.6 Heterogeneity and OSI

There are three levels of heterogeneity in computer networks:[24]

■ The hardware and software that form the systems to be interconnected
■ The physical media and basic protocols of the subnetworks to which the different systems are connected
■ The higher-level protocols used for end-to-end communication among the systems

The OSI model at the higher levels does not accommodate heterogeneity.[24] The real systems, which consist of very different hardware and run very different operating systems, are expected to support the same services and protocols. Particularly, interoperability between networks and systems employing different non-OSI protocols at the network and upper layers is a major problem in the OSI model. The heterogeneity of the real systems has caused not only the addition of proper mechanisms into OSI standards, but also the emergence of different types of services (different modes and functional subsets) and protocols (different classes and options) for a single layer. For example, network and transport layers provide different services and classes to deal with the problem of heterogeneity. Five different classes of the transport protocol have been defined by the transport layer standard to accommodate networks with variable degrees of reliability; the network layer can provide either a connectionless mode or connection-oriented mode service.

This section discusses the problem of interconnecting systems and networks that support different communication architectures. It particularly investigates the important case of non-OSI domains interconnected with and via OSI domains. In LANs, nonstandard protocols are often used to minimize delays in interprocess communication between the modes of a distributed system. Such an environment with its special protocols can be accommodated in the OSI framework as a distributed end system. Gateways can be used to connect such an OSI domain and a non-OSI domain. Gateways supporting such a service must at least perform two basic functions: protocol conversion and interdomain addressing. Protocol conversion can be performed in two ways: service interface mapping and protocol flow mapping. In service interface mapping the conversion takes place between functionally compatible primitives or sequences of primitives offered by the X layer service of architecture AA and the Y layer service of architecture BB. The conversion takes place at the level of the individual protocol data units (PDUs) or sequence of PDUs of the X layer protocol of architecture AA and the Y layer protocol of architecture BB. Interdomain addressing functions enable communication entities in one domain to refer to communication entities in another domain across one or more gateways. Unless different domains use the same global addressing scheme, the interconnecting gateways are required to perform address conversion to enable end-to-end addressability. There are three basic address conversion ways: extended address structure, address encapsulation, and address mapping. The first method consists of adding some type of addressing domain identifier to existing addresses in each of the interconnected domains; this identifier is then used for interdomain routing, while the original addresses are used for routing within each domain. The drawback of this approach is that it typically needs modifications to the original protocols and formats to accommodate extended addresses and allows interdomain routing. This scheme does not accommodate heterogeneity, but overcomes it by imposing a global addressing scheme. In address encapsulation and address mapping, heterogeneity is accommodated at the cost of placing the burden of interdomain routing on end nodes. They must address their traffic directly to appropriate gateways for it to be routed to other domains. The address conversion mechanism used by one subnet is independent of that adopted by other subnets. In case of address encapsulation, the foreign address of a remote entity is encapsulated in a field of a local address denoting the next gateway on the interdomain route; it remains uninterrupted all the way to that gateway, where it is extracted and used onward as it is. When a message is to be routed across multiple domains, its destination address is built through successive encapsulations.

Address encapsulation has two problems. First, it will be possible for the users of each domain to manipulate the formats and meanings of addresses of other domains. This creates security problems. Second, because address fields often are limited in size and content, certain nested addresses may be too long or otherwise unacceptable to be encapsulated locally.

Address mapping allocates a portion of the local address space, corresponding to one or more subfields of local address format, to refer to entities in remote domains across a gateway denoted by the rest of the address. The local addresses representing remote entities are usually called aliases or proxies. When a gateway receives a message addressed to a proxy, it replaces the proxy with the address by which the respective remote entity is known in the domain on the other side. This may be the actual address of the remote entity or another proxy if the message has to be routed across another intermediate domain. This scheme offers uniform interdomain addressing within each domain. In the distributed end system example mentioned above, when the requests for external services are simply exported to a server via an interprocess communication (IPC) facility, the caller could specify the actual OSI address through encapsulation. When a transport-level gateway is provided, address mapping is most likely used. For the interconnection of domains with different architectures, address mapping is often the preferred solution.

1.4   The Emergence of Middleware Architectures

The term *middleware* refers to the software layer between the operating system and the distributed applications that interacts via the network, as shown in Figure 1.5.

It uses basic communication protocols based on messages between processes to provide higher-level abstractions. This software infrastructure facilitates the interaction among distributed software modules. A middleware layer seeks to hide the underlying networked environment complexity by insulating applications from explicit protocol handling, disjoint memories, data replication, network faults, and parallelism. Additionally,

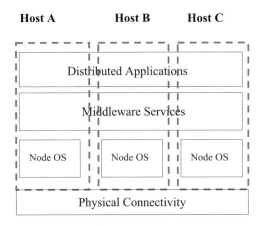

Figure 1.5   Middleware layer position.

middleware architectures hide the heterogeneity of computer architectures, operating systems, programming languages, and networking technologies to facilitate application programming and management. There are two important factors that facilitated the introduction and widespread implementation of middleware: (1) the provision of transparency, openness, and scalability due to the popularity of distributed systems and (2) the use of different types of operating systems, programming languages, and computer architectures. As the demand and use of these service requirements increased, OSI and TCP/IP models were under pressure for modification in their protocol hierarchy. Hence, the middleware layer replaced the session and presentation layers in the OSI model and was introduced between the application layer and transport layer in the TCP/IP stack. The middleware consists of application-independent protocols that do not belong to the lower layers. The functionality of the session and presentation was transferred to other layers in the protocol stack. Middleware communication protocols support high-level communication services. These protocols allow a process to call a procedure or invoke an object on a remote machine through a transparent mechanism. In remote procedure call (RPC) the machine that calls a procedure cannot differentiate whether the procedure runs in the same process on the same machine or in a different process on a different machine. Similarly, in Remote Method Invocation (RMI) the object doing the invocation cannot distinguish whether the object it invokes is local or remote and does not need to know its location. The other examples of protocols that support a variety of middleware services are authentication and authorization protocols; they are not closely tied to any particular application. Commit protocols, on the other hand, allow that either all the processes in a group carry out a particular operation or that the operation is not carried out at all. Commit protocols are used for reliable multicasting applications. Multicasting is not supported by the OSI model. Middleware designs include QoS management and information security. QoS management at the middleware and application level attempts to control attributes such as response time, data accuracy, consistency, and security level.

1.5 The TCP/IP Reference Model

The TCP/IP model originated from the research and development work on ARPANet. ARPANet was an experimental packet-switching network funded by the Defense Advanced Research Projects Agency (DARPA).[3] This model later became known as the TCP/IP reference model or TCP/IP suite. Figure 1.6 shows the TCP/IP reference model.

It consists of five layers:

1. Application layer
2. Transport layer
3. Internet layer
4. Network layer
5. Physical layer

The physical layer is the same as that of the OSI model.

1.5.1 The Network Layer

The network layer provides the means for a host to access the network. The Internet connects heterogeneous networks as they are and utilizes them in an interconnected system of networks to achieve end-to-end

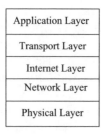

Figure 1.6   The TCP/IP reference model.

communication capability. Hence, the interface to the network and the performance offered by the network are key areas of attention for the network designers. The physical part is concerned with the physical interface between a host and the network. It specifies voltage levels, data rates, etc. As mentioned already, the Internet connects different networks, e.g., packet switching (ATM), LANs (Ethernet), and other types. The network layer provides access and routing independence in such a manner that they are transparent to the end users, i.e., they do not notice any change in their communication.

1.5.2 The Internet Layer

The session and presentation layers are merged in the application and transport layers in the TCP/IP model. The Internet layer is responsible for routing of data packets across multiple networks. A router is an internetworking device that forwards (routes) packets from one network to the other between source and destination. This layer implements the Internet Protocol (IP) to provide this functionality. There are many routing protocols. The packets in this layer are called IP datagrams. Figure 1.7 shows the IP datagram format.

The IP datagram is a variable size packet consisting of a header and data. The header is 20 to 60 bytes in length and consists of routing information. A brief description of each field is as follows:[3]

> **Version (VER)**: It defines the version of the IP. The current version is IPv4. Work is in progress on IPv6.
>
> **Header length (HLEN)**: This field (4 bits) specifies the length of the datagram header. The minimum value of this field is 5, which means a minimum header length of 20 bytes. The maximum value is 15, which indicates a value of 60 bytes.

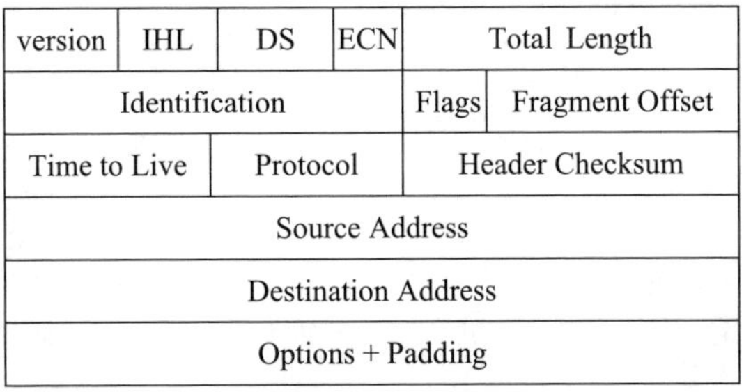

version	IHL	DS	ECN	Total Length	
Identification				Flags	Fragment Offset
Time to Live		Protocol		Header Checksum	
Source Address					
Destination Address					
Options + Padding					

Figure 1.7   The IP datagram format.

Differentiated services (DS): This field (8 bits) specifies class-based quality of service parameters.

Total length: It defines the total length of the IP datagram (header and data) in bytes. The field length is 16 bits; hence, the total length of the IP datagram becomes 65,535 bytes, of which 20 to 60 bytes are the header bytes and the remaining are data.

Identification: This field is 16 bits and identifies a datagram coming from the source host. It is basically a unique sequence number that helps in reassembling the datagram at the destination. All the fragments of a datagram have the same identification value.

Flags: The value of this field is 3 bits. The first bit is reserved. The other bit is called the do not fragment bit. If this bit is set, fragmentation is not allowed. If the size of the datagram is too large and it cannot be transported through the physical network, this datagram will be discarded and an error message generated. The third bit is called the more fragment bit. If this bit is set, it indicates that the datagram is not the last fragment and there are more fragments to follow.

Fragmentation offset: Fragmentation offset of points in the sequence where this fragment fits in. This field is 13 bits.

Time to live: This field tells us about the duration a datagram will remain in the Internet. It is a counter that controls the maximum number of hops visited by the datagram. Each time a datagram passes through a router, the value of this counter is decremented by 1 by the router. If this value becomes zero, the datagram is discarded. The idea is to stop a lost datagram from moving for long times in the Internet. The length of this field is 8 bits.

Protocol: The protocol field specifies the transport protocol to transfer datagrams. The possibilities are TCP and UDP.

Header checksum: The header of the IP datagram changes each time it visits a router. The data part does not change. Hence, the header checksum is recomputed at each router. It is a 16-bit field.

Source address: This is a 32-bit field. It specifies the IP address of the source.

Destination address: This is a 32-bit field. It indicates the IP address of the destination.

Options: This field is reserved for future enhancements in subsequent versions of the IP. This is not an essential part of datagram header processing. That is, datagrams can be successfully transported without it.

1.5.3 *The Transport Layer*

The transport layer is the next layer above the Internet layer. The Transmission Control Protocol (TCP) uses end-to-end mechanisms to ensure reliable, ordered delivery of packets over the connectionless IP layer. It uses flow control, positive acknowledgments with timers, retransmissions, and sequence numbers to achieve reliability across the network. The TCP fragments the data coming into segments and passes these to the Internet layer. There is no provision of flow control in the IP layer. The routers and gateways do not control the flows, and they are unaware of the relation between one message and another. The TCP performs end-to-end flow control by using windows on a per logical connection basis. The segments are reassembled at the destination by the TCP. The segments in TCP are numbered to guarantee ordered delivery at the destination.

1.5.4 *The Application Layer*

The session and presentation layers are not present in the TCP/IP model. They are actually merged into the application and the transport layers as shown in the Figure 1.6. The application layer consists of user-level protocols; e.g., to access Web applications, it uses HTTP; to support electronic mail, SMTP; and to transfer files, FTP.

1.6 Comparison of OSI and TCP/IP Models

The clear difference between the two models is the number of layers. The OSI model has seven layers and TCP/IP has five. The session and presentation layers are merged into application and transport layers in the TCP/IP model.

Both models have transport, application, and physical layers, but other layers are different. In December 1978, the U.S. Department of Defense (DoD) recognized the Transmission Control Protocol (TCP) and the Internet Protocol as official DoD standards to provide secure and reliable communications to the U.S. military. TCP/IP has been in use in military and civilian applications (DoD and ARPANet) for many years. OSI, on the other hand, was adopted by NATO. There has been substantial work done concerning interoperability issues between OSI and TCP/IP.

The TCP standard corresponds roughly to the ISO Class 4 Transport Protocol, and the ISO Transport Protocol Specification was accepted as a Draft International Standard in 1984.[17,18]

1.6.1 Comparison between the TCP/IP and OSI Transport Service Functions

1.6.1.1 Connection Establishment

There are numerous differences between TCP/IP and OSI transport service functions for connection establishment.

> **OSI**: Simultaneous T-CONNECT requests at the two transport service access points (TSAPs) are handled independently by the transport service (TS) provider. Simultaneous T-CONNECT requests typically result in a corresponding number of transport connections.[17,18]
>
> **TCP**: On the other hand, simultaneous T-CONNECT requests at the two TSAPs result in establishment of one transport connection. In a TCP transport connection establishment, the initiator of the establishment fully specifies the global identity of the transport connection wanted, whereas in the OSI transport connection establishment, only the two TSAPs through which the connection is wanted need to be specified.

1.6.1.2 Called Address

The called address parameter conveys the address of the TSAP to which the transport connection is to be established. For TCP, the Internet address plus the port address (the remote socket) have to be used. The TCP port address in the OSI consists of a T-suffix and a transport connection (TC) reference, where the T-suffix identifies the TSAP and the TC reference identifies the transport connection within the TSAP.

Interoperability over the subnet boundaries is not required in layers 1 to 3 of the OSI model. This is due to the fact that these layers mainly deal with node-to-node and internetworking functions. The Internet layer in the TCP/IP suite addresses these issues by assigning IP addresses. The datagrams are then routed through different subnets.

1.6.1.3 Expedited Data Option

> **OSI**: The expedited data option parameter shows whether the expedited data option will be available on the transport connection.
> **TCP**: No expedited data option. Urgent data can be signaled by the use of the intrinsic TCP urgent function.

1.6.1.4 TS User Data

OSI: Limited-length transport protocol data units (TPDUs) are transferred.
TCP: The TCP PUSH function could be used to separate transport service data units (TSDUs) from each other. Data is transferred through a TCP three-way handshake.

1.6.1.5 Data Transfer

There are two modes of data transfer: (1) normal data transfer service and (2) expedited data transfer service. OSI uses a separate flow control for the transfer of expedited TSDUs. TCP has no expedited data transfer option. However, the urgent flag in the header indicates the arrival of TCP urgent data. The receiver takes action on the basis of this urgent flag information. This mode is full duplex.

1.6.1.6 Connection Release Phase

Connection release is performed by the session layer in the OSI model. Its details can be found in the session service definition.

1.6.1.6.1 Orderly Release

Orderly release is available in TCP as the TCP close service. The service is released after all in-transit data has been delivered and accepted. This step is necessary to make sure that all outstanding data is transmitted and received by the sender and the receiver. This service can be initiated by a service user at any time, regardless of the current TC phase.

1.6.1.6.2 Abrupt Release

Abrupt release is allowed at any time regardless of the current TC phase. The delivery of data is not guaranteed once the release phase has started.
   The release can be initiated by:

1. Either or both of the TS users, to release an established TC
2. The TS provider to release an established TC
3. Either or both of the TS users, to finish TC establishment
4. The TS provider, to indicate its inability to establish a requested TC

The OSI user data parameter is used only when the disconnect is user initiated. It is not guaranteed that user data issued in the T-DISCONNECT request by one service user will reach the remote service user.

1.7 Standards Organizations

1.7.1 International Organization for Standardization (ISO)

ISO was founded in 1946.[12] Its members are drawn from the standards creation committees of various governments in the world. ISO is involved in developing various standards in different fields, including science, technology, environment, etc. It is a network of the national standards institutes of 156 countries, on the basis of one member per country, with a central secretariat in Geneva, Switzerland, that coordinates the organization.

In 1946, delegates from 25 countries met in London and decided to create a new international organization, of which the object would be "to facilitate the international coordination and unification of industrial standards." The new organization, ISO, officially began operations on February 23, 1947. Since 1947, ISO has published more than 15,000 International Standards.[12] ISO is developing standards for traditional activities, such as agriculture and construction, from mechanical engineering and medical devices to the latest information technology developments, such as the digital coding of audiovisual signals for multimedia applications.

ISO collaborates with its partners in international standardization, the International Electrotechnical Commission (IEC) and International Telecommunications Union (ITU).[25] These organizations have formed the World Standards Cooperation to better coordinate their activities for the implementation of international standards.

ISO standards are developed by technical committees comprising experts from the industrial, technical, and business sectors. These experts may be joined by others with relevant knowledge, such as representatives of government agencies, testing laboratories, consumer associations, environmentalists, academicians etc. The experts participate as national delegations, chosen by the ISO national member institute for the country concerned. These delegations are required to represent the views of not just the organizations in which their participating experts work, but other stakeholders too. According to ISO rules, the member institute is expected to take account of the views of the range of parties interested in the standard under development and to present a consolidated, national consensus position to the technical committee.

1.7.2 International Telecommunications Union (ITU)

The ITU has played a major role in creating standards for the majority of the world's major telecommunications technologies since its creation in 1865.[25] Presently, the ITU is responsible for the regulation, standardization, coordination, and development of international telecommunications standards. There are three main sectors of ITU: ITU-R (radio communications sector), ITU-T (telecommunications standardization sector), and ITU-D (development sector). ITU-R is responsible for allocating radio frequencies across the world to different groups. ITU-T is responsible for developing standards for telephone and data communications systems. In 1956, the International Telegraph Consultative Committee (CCIT) and the International Telephone Consultative Committee (CCIF) were merged to form the International Telephone and Telegraph Consultative Committee (CCITT), to respond more effectively to the requirements generated by the development of these two types of communication. CCITT floated recommendations in the field of telecommunications and data communications. On March 1, 1993, CCITT was renamed the International Telecommunications Standards Sector (ITU-T). The members of ITU-T are from national governments, companies (AT&T, Vodafone, Cisco, Nortel, Compaq, Sun, Intel, Motorola, Sony, and many more), and regulatory agencies that watch over the telecommunications business, e.g., the U.S. Federal Communications Commission (FCC). Standardization issues are decided by teams of experts. For example, all past modem modulation standards have been generated by these teams. Until mid-1997, 56 Kbps was studied by ITU as well as by a committee from the U.S. Telecommunications Industry Association (TIA). These two organizations joined their efforts and worked together for several months to arrive at the standard V.90. Since its start, ITU-T has produced around 3000 recommendations.[25] As standards are becoming more and more important in this global society, the ITU-U aims to promote and extend the benefits of new telecommunication technologies to the world's inhabitants.

1.7.3 American National Standards Institute (ANSI)

Founded in 1918, ANSI is a private, nonprofit, and nongovernmental association.[26] Its members are vendors, network providers, and other technical units. All ANSI-accredited standards developers follow the globally accepted principles of standardization implemented by well-recognized international standards bodies such as the International Telecommunications Union (ITU), International Organization for Standardization (ISO), and International Electrotechnical Commission (IEC). The institute represents the interests of its nearly 1000 company, organization, government agency, institution, and international members through its office in New York City and its headquarters in Washington, D.C. ANSI facilitates the development of American National Standards (ANSs) by accrediting the procedures of standards developing organizations (SDOs). These groups work cooperatively to develop voluntary national consensus standards. Accreditation by ANSI signifies that the procedures used by the standards developing organizations in connection with the development of American National Standards meet the institute's essential requirements of openness, balance, and consensus. To maintain ANSI accreditation, standards developers are required to consistently adhere to a set of requirements or procedures known as the ANSI Essential Requirements, which govern the consensus development process.

1.7.4 Institute of Electrical and Electronics Engineers (IEEE)

The IEEE is the largest international engineering organization.[27] It has more than 365,000 members in over 150 countries and more than 27,000 society members, 307 sections, and 1446 chapters.[27] It has 39 technical societies and 3 technical councils covering the range of electrical engineering and information technologies. The IEEE originated from its two predecessors: the American Institute of Electrical Engineers (AIEE) and the Institute of Radio Engineers (IRE). AIEE was founded in 1884 and IRE in 1912. Many of the original members of the IRE were members of the AIEE, and both organizations continued to have members in common until they merged to form the IEEE in 1963. The IEEE is a leading organization in technical fields ranging from computer engineering, biomedical technology, and telecommunications to electric power, aerospace engineering, consumer electronics, and many more. The IEEE publishes many journals and organizes several conferences each year around the world. It also develops standards in the field of electrical engineering and computer science. It has nearly 900 active standards, with almost 500 under development. It has developed standards for LANs, such as IEEE 802.3 for Ethernet LANs and IEEE 802.11 for wireless LANs (WLANs), components and materials, information technology, electromagnetics, power electronics, medical devices, transportation technology, and many more.

1.7.5 Electronic Industries Association (EIA)

The EIA is a nonprofit organization.[28] In the field of electronic communication it has made important contributions by specifying physical connection interfaces and electronic signaling standards. For example,

EIA/TIA/RS-422 provides full-duplex asynchronous point-to-point communication via two twisted pairs.

1.7.6 Internet Engineering Task Force (IETF)

The IETF is a large, open international community of network designers, operators, vendors, and researchers concerned with the evolution and smooth operation of the Internet.[29] It is open to any interested individual.

The IETF works through its technical working groups, which are organized by topic into several areas (e.g., routing, transport, security, etc.). Much of the work is handled via mailing lists. The IETF holds meetings three times a year.

The IETF working groups are grouped into areas and managed by area directors (ADs). The ADs are members of the Internet Engineering Steering Group (IESG). The Internet Architecture Board (IAB) provides architectural supervision. The IAB also arbitrates the disputes when someone complains that the IESG has failed. The IAB and IESG are chartered by the Internet Society (ISOC) for these purposes.

The Internet Assigned Numbers Authority (IANA) is the central coordinator for the assignment of unique parameter values (e.g., port numbers) for Internet protocols. The IANA is chartered by ISOC to assign and coordinate the use of numerous Internet protocol parameters.

1.8 Summary

Traditional networking protocols were built for largely non-real-time data with very few burst requirements. The protocol stack at a network node is fixed, and the network nodes only manipulate protocols, up to the network layer. New protocols such as the Real-Time Transfer Protocol (RTP) and Hypertext Transfer Protocol (HTTP) enable the network to transport other types of application data such as real-time and multimedia data. Such protocols cater to specific QoS demands of the application data. But transporting these new data types over a legacy network requires us to transform the new type of data into the type of data carried by the network. Transforming the data to fit legacy protocol requirements prevents one from understanding the transformed protocol. For example, embedding a Motion Picture Experts Group (MPEG) frame in Multipurpose Internet Mail Extensions (MIME) format prevents us from easily recognizing an I, P, or B frame. This prevents the network from taking suitable action on the MPEG frame during times of congestion.

Introducing new protocols in the current infrastructure is a difficult and time-consuming process. A committee has to agree on the definition of a new protocol. This involves agreeing on a structure, states, algorithms, and functions for the protocol. The time from conceptualization of a protocol to its actual deployment in the network is usually extraordinarily long.

Traditional protocol frameworks use layering as a composition mechanism. Protocols in one layer of the stack cannot guarantee the properties of the layers underneath it. Each protocol layer is treated like a black box, and there is no mechanism to identify if functional redundancies occur in the stack. Sometimes protocols in different layers of the same stack need to share information. For example, TCP calculates a checksum over the TCP message and the IP header,[2] but in doing so, it violates modularity of the layering model because it needs information from the IP header that it gets by directly accessing the IP header. Furthermore, layering hides functionality, which can introduce redundancy in the protocol stack.

The solution to these problems is active and programmable networks. For example, an advantage of active networking is that it enables application-specific processing through the dynamic deployment of user-defined functionality at the network nodes. This functionality can be tailored to the data type being transported. The challenge is to identify an approach that enables creation of flexible, custom, complete, and correct programs, which can be then injected into the network.

Exercises

1. Discuss the importance of protocols and standards for network communications.
2. Discuss the major constraints imposed by OSI and TCP/IP models. Compare these two models.
3. Why are reference models layered?
4. Does the OSI model cater to the heterogeneity of distributed systems?
5. How does middleware cater to heterogeneous environments?

References

1. D.L. Tennenhouse et al., A survey of active network research, *IEEE Commun. Mag.*, 35:1, 80–86, 1997.
2. S.A. Hussain and A. Marshall, An agent based control mechanism for WFQ in IP networks, *Control Eng. Pract.*, 11:10, 1143–1151, 2003 (special issue on control methods for telecommunication networks).

3. B.M. Leiner, R. Cole, J. Postel, and D. Mills, The DARPA Internet protocol suite, *IEEE Commun. Mag.*, 23:3, 29–34, 1985.

4. www.uucp.org and http://www.walthowe.com/navnet/history.

5. www.cse.wustl.edu.

6. http://archive.dante.net/europanet.html.

7. P.E. Green, Jr., An introduction to network architectures, *IEEE Trans. Commun.*, 28:4, 413–424, 1980.

8. J.B. Postel, Internetwork protocol approaches, *IEEE Trans. Commun.*, 28:4, 606–611, 1980.

9. D.D. Clark and D.L. Tennenhouse, *Architectural Considerations for a New Generation of Protocols*, ACM, 20:4, 200–208, 1990.

10. D.D. Clark, V. Jacobson, J. Romkey, and H. Salwen, An analysis of TCP processing overhead, *IEEE Commun. Mag.*, 23–29, 1989.

11. Y. Yemini, The OSI network management model, *IEEE Commun. Mag.*, 1993.

12. H. Zimmermann, OSI reference model: the ISO Model Architecture for Open Systems Interconnection, *IEEE Trans. Commun.*, 28:4, 425–432, 1980.

13. P.S Kritzinger, A performance model of the OSI Communication Architecture, *IEEE Trans. Commun.*, 34:6, 554–563, 1986.

14. R. Popescu-Zeletin, Some critical considerations on the ISO/OSI RM from a network implementation point of view, *ACMS/GCOMM Symposium*, 188–194, 1984.

15. R. Desjardins, *ISO Open Systems Interconnection Standardization Status Report*, ACM, 1983.

16. R. Popescu-Zeletin, *Implementing the ISO-OSI Reference Model*, ACM, 1983.

17. M.T. Rose, Transition and coexistence strategies from TCP/IP to OSI, *IEEE J. Selected Areas Commun.*, 8:1, 57–66, 1990.

18. I. Groenbaek, Conversion between the TCP and ISO transport protocols as a method of achieving interoperability between data communication systems, *IEEE J. Selected Areas Commun.*, 4, 288–296, 1986.

19. H.V. Bertine, Physical level protocols, *IEEE Trans. Commun.*, 28:4, 433–444, 1980.

20. S. Keshav, *An Engineering Approach to Computer Networking, ATM Networks, the Internet, and the Telephone Networks*, Pearson Education, Singapore, 1997.

21. A.E. Conway, A perspective on analytical performance evaluation of multi-layered protocol architectures, *IEEE J. Selected Areas Commun.*, 9, 1993.

22. C. Tschudin, *Flexible Protocol Stacks*, ACM, 197–205, 1991.

23. M.B. Abbot and L.L. Peterson, Increasing network throughput by integrating protocol layers, *IEEE Trans. Networking*, 1:5, 600–610, 1993.

24. L. Svobodova, P.A. Janson, and E. Mumprecht, Heterogeneity and OSI, *IEEE J. Selected Areas Commun.*, 8:1, 67–79, 1990.

25. www.ITU.org.

26. www.ansi.org.

27. www.ieee.org.

28. www.eia.org.

29. www.ietf.org.

Chapter 2

Architecture of Active and Programmable Networks

2.1 Introduction

Packet-switched networks enable sharing of transmission facilities so that packets may be efficiently transferred among the network nodes. Present-day networks perform only the processing necessary to forward packets toward their destination. Active and programmable networks represent a quantum jump in the research of packet-switched networks by providing a programmable interface in the intermediate network nodes. Active and programmable networks allow injection of customized code within routers and switches by active packets or open signaling.[1] They support dynamic modification of network behavior as seen by a user and a service provider.

In 1995, the Defense Advanced Research Projects Agency (DARPA)[1] floated the concept of active networking with the notion of challenging the current problems in traditional networks, such as the difficulty of integrating new technologies, standards, and services into the shared network infrastructure. Later on, two schools of thought emerged on how to make networks programmable: the Programmable Switch Network Community and the Active Networks Group.[2] The Programmable Switch Network Community pursued the OPENSIG (open signaling) approach. The Programmable Switch Network Community is of the view that by modeling communication hardware using a set of open programmable network interfaces, open access to switches and routers can be provided.[2] They maintain that by opening up the switches in this way, enhanced control over the network objects and development of new and distinct architectures and services, e.g., virtual networking, can be realized. OPEN-SIG takes a telecommunications approach to the problem of making networks programmable. In this approach, physical network devices are abstracted as distributed computing objects, e.g., virtual switches, switchlets, and virtual base stations, with well-defined open programmable interfaces. These open interfaces allow service providers to manipulate states of networks, using middleware toolkits, e.g., Common Object Request Broker Architecture (CORBA), to construct and manage new network services.[2]

The active network group, on the other hand, proposed dynamic deployment of new services on the fly, mainly within the confines of Internet Protocol (IP) networks. Active networking allows intermediate nodes in the network to be programmed dynamically to be user and application specific by active packets (capsules). The end users can therefore inject customized programs into the network elements and modify the behavior of the network. In active networks, packets represent the main vehicle for program delivery, control, and service creation. At one extreme, a single packet could boot a complete software environment seen by all packets arriving at the node. At the other extreme, a single packet could modify the behavior seen by that packet. Active networks allow customization of network services at the packet transport level rather than through a programmable control plane.[2] Theoretically, this approach can be more dynamic than quasi-static programmable networks, because it might be possible to customize network services at the granularity of individual packets, but at the cost of more complexity to the programming model.

2.2 Quality of Service Technologies for IP Networks

Broadband IP networks support a variety of applications, ranging from voice and data to multimedia applications. The service quality requirements of these applications vary over a wide range: some are sensitive to delays experienced in the communication network, others to loss rates, and others to delay variation. Moreover, because different applications have widely different traffic characteristics, the network must be able to satisfy the quality of service (QoS) requirements of each of these applications. Hence, the concept of QoS support is an important issue for equipment vendors, services providers, network providers, and network operators. Along with signaling, routing, and policing, QoS support has two other implementation domains: (1) congestion management and (2) congestion avoidance. Congestion management includes packet-scheduling mechanisms, and congestion avoidance consists of packet-dropping policies.

Because QoS requirements of applications vary over time, an appropriate QoS technologies support framework is needed for specifying application requirements and translating those requirements into network resources. Some of the most recent architectures and frameworks for implementing QoS support for IP-based networks have been proposed in Campbell et al.[2] Additionally, standard bodies such as the Internet Engineering Task Force (IETF)[3] are also addressing QoS problems. Protocols such as the Resource Reservation Protocol (RSVP) and Differentiated Services (DiffServ) have been introduced by the IETF.

QoS is the ability of a network element (e.g., switch or router) to give some level of assurance to its traffic such that service requirements can be satisfied. From the end users' point of view, it means providing consistent predictable service delivery that satisfies users' applications requirements. It relies heavily on the underlying network technologies and involves the cooperation of all network elements involved in a connection. The primary goals of QoS include dedicated bandwidth, controlled jitter (delay variation), end-to-end latency (required by some real-time and interactive traffic), and improved loss characteristics.[4,5]

The QoS requirements for best-effort applications and guaranteed-service applications are different. The performance requirements for best-effort applications are flexible in the sense that they adapt to the resources available, for example, File Transfer Protocol (FTP) over Transmission Control Protocol (TCP). They work even with degraded performance as the available bandwidth decreases and end-to-end delay increases. Best-effort applications do not require the network to reserve resources for a connection. They are more sensitive to packet loss than to packet latency.

Guaranteed-service applications, on the other hand, require a guarantee of QoS from the network. These applications require the network to reserve resources. Each customer makes unique demands for some level of service predictability, even in the presence of transient congestion due to other traffic traversing the network. These demands include low end-to-end delay or latency, controlled jitter for real-time applications such as Voice-over-IP, and videoconferencing and low packet loss for non-real-time applications (FTP, e-mail, etc.). These QoS requirements lead directly to the dynamics of queue utilization and queue management within each router. Hence, packet classification, queuing, and scheduling are the most important network resources to be configured for providing QoS. Packet-scheduling disciplines play a critical role in controlling the interactions

Table 2.1   IP End-to-End QoS Technologies

Description	Technology	Functional Domain
Out-of-band signaling, in-band signaling	RSVP (InterServ), DS byte (DiffServ)	Signaling
Traffic policing	CAR	Policing and classification
Weighted fair queuing, class-based queuing (scheduling mechanisms)	WFQ, CBQ	Congestion management
Random early detection, weighted random early detection (packet-dropping policies)	RED, WRED	Congestion avoidance
MPLS DiffServ, IP + ATM QoS integration	MPLS	Routing + switching

among different traffic streams and different classes of traffic. Besides deciding the order in which requests are serviced, a scheduler allocates different delays to different users by the choice of service order. It also allocates different loss rates to different users by its choice of which requests to drop. There are many schedulable resources; however, the most important ones are the bandwidth available over a link and the allocations of buffers in a multiplexing point according to the class of traffic (classification). Table 2.1 gives a summary of IP end-to-end QoS framework technologies.

Integrated Services (IntServ)[6,7] supports per flow QoS guarantee. In this architecture, each traffic is differentiated according to its destination address, destination port, and protocol type. It supports two service classes in addition to best-effort service: guaranteed service (GS) and control load (CL). GS provides a strict guarantee for both throughput and worst-case end-to-end delay bound.[8] It makes sure that the IP packets are served within an agreed upon time and will not be discarded due to queue overflow. An application that subscribes to CL receives quality that resembles a best-effort service under lightly loaded conditions.[9] It keeps the number of lost or dropped packets to a minimum.

The Resource Reservation Protocol (RSVP)[7] is the signaling protocol for IntServ for the configuration of connections. It was developed to support the signaling requirements for both end-to-end unicast and multicast flows with QoS requirements. To initiate reservation, the source sends a PATH (path establishment) message downstream to the receiver.

The receiver then responds with an RESV (reservation) message upstream toward the sender following the exact route taken by the PATH message. The PATH message installs a path state in each router along its path so that the RESV message can flow the same route in the reverse direction. The PATH message also carries descriptions of the sender's traffic characteristics and network characteristics through SENDER_TSPEC and ADSPEC (advertising), respectively. Each reservation only has a limited lifetime. Hence, the PATH and RESV messages must be retransmitted periodically during each session to keep the reservation states alive. As a result of this property, RSVP leaves a large amount of soft-state information in the core routers. This creates scaling problems.

Differentiated Services (DiffServ),[10] on the other hand, aims to provide a simpler functionality than IntServ. This approach does not keep end-to-end flow states within the core routers. The IP packets entering a DiffServ network are aggregated and treated according to their traffic classes and not as individual flows. The packets are classified using a 6-bit DiffServ Code Point (DSCP), which subsequently defines the traffic conditioning actions such as marking, metering, shaping, or policing (dropping). These actions are normally enforced by the ingress routers at the edge of the network. The packet handling is performed on a hop-by-hop basis or per hop behaviors (PHBs). PHBs are the forwarding behaviors applied to aggregates at core routers.

The committed access rate (CAR) controls the maximum rate for traffic transmitted or received on an interface. The CAR is often configured on interfaces at the edge of a network to limit traffic into or out of the network. Traffic that falls within the rate parameters is transmitted, while packets that exceed the acceptable amount of traffic are dropped or transmitted with a different priority. It also classifies the packets through IP precedence and QoS group setting.

Multiprotocol Label Switching (MPLS) is a traffic engineering approach in which a router determines the next hop in a packet's path without looking at the packet header or referring to routing lookup tables.[11] The MPLS assigns short fixed-length labels to packets as they enter the network. The network uses these labels to make forwarding decisions, usually without looking at the original packet headers.

Random early detection (RED) and weighted random early detection (WRED) are congestion avoidance mechanisms. RED[12] detects incipient congestion by computing the average queue size; when it exceeds a preset threshold of average queue size, arriving packets are dropped with a certain probability that is a function of the average queue size. The result of the drop is that the source detects the dropped traffic and slows its transmission. RED is primarily designed to work with the Transmission Control Protocol/Internet Protocol (TCP/IP). WRED[13] selectively discards lower-priority traffic when the interface starts to get congested and provides differentiated performance characteristics for different applications.

2.3 Quality of Service Parameters

The important QoS parameters for efficient performance of a network are described in this section.

2.3.1 End-to-End Latency

End-to-end latency is an important parameter for real-time applications. The end-to-end latency experienced by a packet is a combination of the transmission delays across each link and the processing delays (which include buffering or queuing) experienced within each router. The delay introduced by link technologies such as the synchronous optical network (SONET) or synchronous digital hierarchy (SDH) or constant bit rate (CBR) Asynchronous Transfer Mode (ATM) virtual circuits is predictable.[13] However, in IP networks the delay is not so predictable. It varies with the variable packet sizes (e.g., FTP, bursty video, etc.) and changing congestion patterns, often varying from one moment to the next even for packets heading for the same destination. The end-to-end delay, for example, for Voice-over-IP (Internet telephony), is contributed by sources such as queuing delay in routers, coding delay (frame processing delay and look-ahead delay), serialization delay, and propagation delay. Higher compression is achieved at the cost of longer delays. Propagation delay for Internet telephony becomes more significant where long distances are involved, e.g., when calls are routed over a satellite link. An end-to-end delay of 150 ms is usually acceptable for Voice-over-IP.[14,15] Voice quality deteriorates significantly above 150 ms.

2.3.2 Delay Jitter (Delay Variation)

The delay jitter for a packet stream is defined to be the absolute difference between the delays between received packets. Delay jitter bound is an important QoS parameter for applications such as videoconferencing, Voice-over-IP, and video on demand. A very important requirement for real-time applications is the preservation of the temporal ordering of packets, i.e., the time between arrivals of packets from the same source.

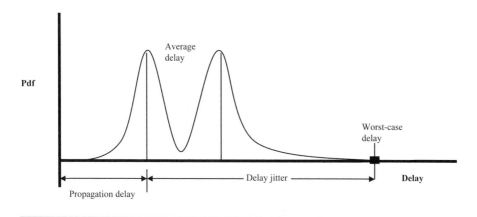

Figure 2.1   End-to-end delay and delay jitter.

If an application sends a stream of regularly spaced packets and they arrive at the far end as bursts of packets clumped together in time, then the network has disrupted the source's temporal ordering. This phenomenon is known as delay jitter.

Delay jitter bound can be as large as the delay bound less the propagation delay, because a packet's queuing delay could be as small as zero and as large as the delay bound less the propagation delay.[16] A delay jitter bound requires that the network bound the difference between the largest and smallest delays received by packets on that connection (bounds the width of the end-to-end delay histogram), as shown in Figure 2.1.[16] For example, playback applications such as video on demand use jitter buffers or playback receivers to eliminate the jitter. If the delay jitter over a connection is bounded, the receiver can eliminate delay jitter by delaying the first packet on that connection by the delay jitter bound, then playing the packets out from the connection a constant time after they were generated. The jitter buffer hold time adds to the overall delay. Therefore, for high jitter, the overall perceived delay is high even if the average delay is low. For example, the overall delay is only 55 ms for a moderate average delay of 50 ms with a 5-ms jitter buffer. In contrast, if the network has a low average delay of 15 ms, but occasionally the packet is delayed by 100 ms (the delay buffer would have to be 100 ms), the overall delay in this case will be 115 ms.[14] Common buffer sizes range from 50 to 100 ms.[14] Figure 2.1 shows the end-to-end delay distribution for a typical connection. In Figure 2.1, the end-to-end delay is bounded by propagation delay on one side and the worst-case delay on the other side.

2.3.3 Packet Loss

Another important issue is packet loss. Given that routers have only finite buffering capacity, a sustained period of congestion may cause the buffers to reach their capacity. When packets arrive to find buffer space exhausted, packets must be discarded until buffer space becomes available. Packet-dropping schemes such as RED and WRED drop packets to relieve network congestion. Applications such as the File Transfer Protocol (FTP) and e-mail can tolerate long queuing delay, but they cannot tolerate high packet losses. The main intention in these applications is the complete transfer of information and not the timing of delivery, whereas in the case of applications such as streaming video and audio, timely delivery of applications is very important. Packets are discarded if they are held too long in the queue, and longer queues can lead to higher end-to-end delays, and hence poor QoS.

2.4 Motivation for Active and Programmable Networks

First, traditional networks have the drawback that intermediate nodes (e.g., routers, switches) are vertically integrated closed systems whose functions are rigidly built into the embedded software and hardware by intermediate node vendors.[17,18] Therefore, the development and deployment of new services in such networks requires a long standardization process instead of a rapid introduction of innovative cost-effective technologies. The range of services provided by these types of networks is also limited because they cannot anticipate and provide support for all future applications. Programmable networks, on the other hand, offer a different approach that enables programming of intermediate nodes in the network. Programmable networks accelerate the pace of innovation by decoupling networks from the underlying hardware and allowing new services to be loaded into the infrastructure on demand. In this way, users could have their own virtual private networks, over which they can exercise their control and authority.

The second factor that motivated research into active networks was the decision to provide a well-designed and architecturally open platform, and the efforts that had already been going on to achieve this. Examples in this context are RSVP, which reserves the bandwidth to ensure that time-sensitive data is delivered in a timely fashion; IP multicast, which reduces the bandwidth needed to communicate from one sender to multiple receivers; and mobile IP, which allows a roaming laptop to be reached at different sites without the need to reconfigure address information. Other examples are firewalls and Web proxies.[17] Firewalls allow implementation of application- and user-specific functions by determining which packet should be passed transparently and which should be blocked. There is always a need to update the firewall resources due to the use of new applications. Active networks can facilitate this process by allowing new applications and injecting the appropriate modules into it. Web proxies deal with the serving and caching of Web pages. There are numerous examples where active Web proxies can help reduce the latencies of the Internet. The traditional Web caches in applications like stock quotes and online auctions are of little help. Most Web caches do not cache quotes because they are dynamic. The other issue pertains to the scalability of storing the number of unique Web pages. Clients in this environment request pages with a list of quotes. With a large number of quotes, the possible number of required Web pages will be huge; this results in very low cache hits. Active networks can improve performance by allowing caching of stock quotes at the network nodes on a per stock name basis, instead of a large number of stocks per page. This increases the cache hit ratio. The second improvement is the client's control over the degree of currency ("updateness") of the stock quotes. For example, some clients may be perfectly satisfied with the quotes that are 16 min out of date, while others may need the latest quotes. The quotes, along with their time stamps, are cached at network nodes as they move from the server to a client. Client requests later on are interpreted at the intermediate nodes, where their local caches are checked for the requested quotes. If the requested quotes are available, the results are forwarded to the clients. If not, the requests are forwarded to the server. In the case of online auctions, the active nodes can filter out the low bids before they reach the server. These bids are delayed because of the network delay. This capability enhances the server throughput when the network is heavily loaded. When the load on the network increases, the server activates filters on the nearby nodes. These filters drop the bids that are lower than the current price and send rejection notifications to the clients. An argument regarding the performance of the above examples can be that it increases the amount of processing within the network. But in fact, it leads to improved overall performance. In the online auction example, there is an increase in the number of successful bids per second. Similarly, in the example of stock quotes, there is an increase in the number of client requests serviced per second. Both of these improvements are brought about by shifting some of the application's functionality to internal nodes of the network. Apparently these improvements come at the cost of increased computational and storage resources in the network. This may slow down traffic through the network. Because active node processing reduces bandwidth utilization in some sections of the network, this will reduce the delays and packet losses of applications. Another important point is the placement of active nodes in the network. In the stock quote example, it is critical to deploy the Web proxies at a place where they can serve the maximum number of clients requests; otherwise, they will not be fully utilized.[17] In the online auction example, active nodes should be near to the clients to send back low bits as quickly as possible, but close enough to the server to get updated price information.[17]

The third motivation was to gain the benefits of being able to put application and network knowledge together in the same place and time. Stock quotes and online auctions are typical examples. In online auctions the server collects and processes client bids for the available items. This server also responds to requests from clients for the current price of an item. Because of the network delay experienced by a packet responding to such a query, its information may be out of date by the time it reaches a client, possibly causing the client to submit a bid too low to beat the current going price. In active networks, when the server senses that it is heavily loaded, it can activate filters in nearby nodes to drop bids lower than the latest bid and periodically update them with the current price of the items by sending active packets.

The final motivation was to provide an integrated mechanism for security, authentication, and monitoring, thus eliminating the need for multiple security/authentication systems that operate independently at each communication layer protocol. Programmable networks can provide a security policy for the network on a per user or per use basis. By using active network acknowledgment (ACK) aggregation and buffering, implosion problems can be avoided and throughput can be increased. Active networks can perform ACK fusion at active nodes between the receivers and the source. ACK fusion consists of sending just one ACK from a given active node toward the source of each ACK received. The new ACK carries the fused information of all n ACKs. Active networks can also perform NACK (negative acknowledgment) filtering at active nodes between the receivers and the source. The active nodes remember the data already requested, and when a NACK is received, data is forwarded only if a different set of data is requested.

In active networking, the processing time at the network nodes is traded for the benefits discussed above. Additional benefits can be in the form of reduced bandwidth demand, better utilization of node resources such as buffers and queues, and application-specific customization.

2.5 The IEEE 1520 Standards Initiative for Programmable Networks

Open programming interfaces define the boundary between the hardware platform and software control to enable an open service creation environment where services may be created faster than the underlying network. Programming interfaces allow service providers and third-party network providers to program equipment from different vendors.

The OPENSIG community has already been working to standardize programmable networks. The IEEE project 1520 (IEEE P1520)[19] on programmable interfaces for networks has followed the OPENSIG approach in an attempt to standardize programmable interfaces for ATM switches, IP routers, and mobile telecommunication networks.

A present-day signaling network is built with a few fixed algorithms and programs known as standard signaling protocols and control programs. Development of new signaling protocols has been a slow and difficult process. This is because the signaling standards have become very complex, and their implementation into the products requires consensus from vendors, which is a long process.

The solution to the above problem was proposed by IEEE P1520 by providing a set of software abstractions of network resources that allowed distributed access to low-level control functionalities of network devices. A reference model was developed by IEEE that partitioned end-user application semantics, value-added services, network-generic services, and hardware and software support. The project supported the idea of establishing programming interfaces between the levels of the reference model. The reference model provided a set of programmable interfaces for distributed access to switching functions by service control entities. It also provided open interfaces between the hardware and operating system and between the operating system and applications. Figure 2.2 shows different levels of the P1520 reference model.

Like the Open System Interconnect (OSI) and TCP/IP models, there are different entities at each level and interfaces between levels. At the value-added services level (VASL) there are algorithms that add value to services provided by the lower services through capabilities such as real-time stream management, synchronization of multimedia, etc. The network-generic services level (NGSL) consists of algorithms that basically deal with the functioning of the network, for example, virtual connection/virtual path (VC/VP) setting algorithms or routing algorithms. At the virtual network device level (VNDL), the entities are logical representations (in terms of objects) of certain state variables of these entities in the VASL. The interfaces at this level are the abstractions of resources of the physical elements at the physical element (PE) level. An example of this level is

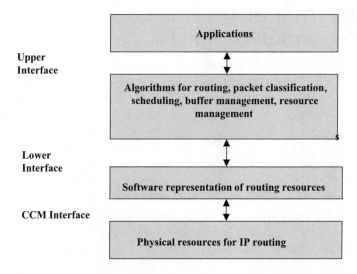

Figure 2.2   Different levels of the P1520 reference model.

the binding interface base (BIB), which is an abstraction of the VC/VP namespace.

The entities at the PE level are physical elements of the network, such as ATM switches and IP routers, multiplexers, cross-connects, etc. These elements can be accessed by means of control protocols, such as the General Switch Management Protocol (GSMP) of Ipsilon. GSMP is a master/slave protocol that allows a node, for example, an ATM switch, to be controlled by an external source to establish and release virtual connections. The ATM switch exchanges GSMP messages through an ATM virtual connection identified at initialization. GSMP consists of five types of messages:[19]

1. Connection management messages to establish and release virtual connections
2. Port management messages to activate or deactivate specific switch ports
3. Configuration messages to discover the capabilities of the switch
4. Statistics messages to request information on VC or port activity
5. Event messages to report alarms

The four types of interfaces identified in this model are:[19]

1. The V interface, which provides access to the value-added services level.

2. The U interface, which provides services from the NGSL to the VASL.

3. The L interface, which provides the application programming interface (API) to directly access and manipulate local network resources. These resources could be VC/VP lookup tables in the case of ATM networks or routing tables for IP forwarding in the case of IP routers.

4. Connection control and management (CCM) interface, which provides access to physical elements.

2.5.1 Programming Interfaces for ATM Networks

The programming interfaces for ATM networks allow specification of QoS requirements and constraints. These generic QoS abstractions introduce programmability into the ATM network by specifying any signaling and control procedure in terms of a sequence of remote operations over these abstractions. These abstractions are specified at the L interface in the P1520 reference model. They represent a subset of the network resources as defined in the L interface of the P1520 reference model and make the ATM network programmable.

There are two broad categories of interfaces. In the first category, two fundamental forms of resources are identified. They are namespace and bandwidth. The namespace resources exist as the virtual path and virtual circuit identifiers (VPIs, VCIs) of an ATM channel, while bandwidth resources appear as the buffer sizes at the multiplexing points of an ATM cross-connect. Interfaces in the second category can be further classified into sub-APIs that support control of multimedia devices, ATM switch fabric and output multiplexers, etc. Control of multimedia devices refers to the APIs to allow the setup and subsequent changing of multimedia device parameters that change the generation or use of multimedia data. These typically include control of sampling parameters, rate, format, encoding scheme, and much more. The control of ATM switch fabric refers to the task of manipulation of VC/VP lookup tables in an ATM switch fabric and the control of scheduling policies at the output ports.

It follows a layered view of a programmable network that is similar to the idea of an operating system.[19] With an analogous software layer over a programmable network, different control architectures, including signaling protocols, can operate on the same network. The lowest layer consists of physical resources for forwarding IP datagrams, e.g., routing tables, buffers, packet classifiers, packet schedulers, and switch fabrics. They may be directly accessed through the connection control and management (CCM) interface. The CCM interface provides access to management protocols to collect state information. Although programming interfaces were initiated for ATM network control rather than IP network control, the need for dynamic modification of policies and configurations within IP routers emerged in the domain of active networks later on.

2.5.2 Programming Interfaces for IP Router Networks

Figure 2.2 shows mapping of the P1520 reference model to IP routers. The programmable interface for a router supports traffic control parameters by dynamically adjusting resource allocations among differentiated service classes, alternative routing algorithms, and discriminant treatment of packets associated with an application or user. P1520 defines programming interfaces for IP routers. In Figure 2.2, several classes of algorithms operate at the NGSL. These classes can be used for routing table lookup and manipulation. These algorithms, for example, are policy-based routing, Differentiated Services scheduling, Resource Reservation Protocol (RSVP), or other flow-based protocols.

The research community has suggested using Multiprotocol Label Switching in a trunk dedicated to a particular service class and RSVP for setting up and maintaining these trunks. The other view is to use RSVP and DiffServ in different parts of the network.[3] RSVP and DiffServ have been proposed as mechanisms for implementing QoS in IP-based networks, which basically provide a best-effort-only service. However, RSVP has scaling problems, and its implementation on nodes that transport a large number of traffic streams will leave a large soft state in the routers that is difficult to scale. Alternatively, DiffServ provides a granular service, which cannot fully optimize network performance. The solution proposed by A. Marshall[20] considers the view that future networks will consist of combinations of these mechanisms: RSVP may be more efficiently employed in the access network, whereas DiffServ will be deployed in the higher-density core network. Efficient implementation of this project would allow the network to carry higher traffic loads with a wider range of traffic profiles.

2.6 Classification of Active and Programmable Networks

2.6.1 Discrete Mechanism Approach

In this mechanism the processing of messages is architecturally separated from the function of injecting the programs into the node, with a separate mechanism for each function.[1] In this approach users send their packets through the programmable nodes. At the nodes, the packet header is examined and an appropriate program is dispatched to operate on its contents. It is possible to arrange for different programs to be executed for different users or applications. This approach of separation of program execution and loading is important when security is required for program downloading or when programs are large. In the Internet, environment code can be injected into the routers. The router will perform authentication and extensive checks on the code. This dynamic code loading into the routers is useful for router extensibility purposes.

2.6.2 Integrated Mechanism or Capsule Approach

In this approach, every capsule or message that passes through the nodes contains a program fragment that may contain embedded data.[1,21] At the nodes, their contents are evaluated and then dispatched to a transient execution environment where they can be safely executed. Framing mechanisms provided by traditional link layer protocols can be used to identify capsule boundaries. The programs can be composed of instructions that perform basic computations on the capsule contents or can also invoke primitives that may provide access to resources external to the transient environment. This approach is also known as *in-band control* in the context of active networks.

Active networks can be tailored to support a variety of traditional functions, such as IP packet forwarding, classification, flows, routing, scheduling, and so on. For example, in pure source routing algorithms each message carries the identities of all the links it travels. The IP mechanisms for determining and naming the links on which outgoing capsules are sent are built in in every node, and individual packets need not know the links they travel. They only have to carry their destination addresses. With the active network approach, capsules can dynamically enumerate and evaluate the paths available at a node without needing detailed knowledge at the time the capsule is composed.

The next issue is the level at which a capsule program can access modules such as routing tables that are located beyond the transient execution environment. There are three ways in which capsule programs could reach beyond the capsule's transient environment:

- Foundation components: Universally, services implemented outside of the capsule.
- Active storage: The ability to modify the state at which the node storage is left at the completion of capsule execution.
- Extensibility: Allowing programs to define new classes and methods.

2.6.2.1 Foundation Components

The aim of implementing foundation components is to provide controlled access to resources external to the transient environment. Capsules will require access to other node-specific information and services such as routing tables and the state of the node's transmission links. There will be two categories of these components; one of them will expose the API to the applications. The other will provide a built-in object-oriented hierarchy that serves as a base for the development of capsule programs. The capsules designed by using these built-in components can perform processing similar to that performed on the header of an IP datagram. Multicasting and other options (in the header of an IP datagram) processing instruction could be included in the capsules that need them. This is in contrast to the traditional IP approach, where the code is fixed and built into the router, whereas in the capsule case, the program/code is flexible and carried with the data. For interoperability with the existing routers, active networks will consist of standardized components that implement the existing Internet protocol types. The implementing legacy nodes will forward the packets using IPv4.

2.6.2.2 Active Storage

Every capsule would include code that attempts to locate and use its flow state at the nodes it travels through. In case a flow capsule has to pass through a node that has no relevant state information, it dynamically generates the required data, uses it for its own functions, and leaves it behind for the later arriving capsules. The network can treat the flows as a soft state, compared to connections. The soft state consists of routing table information in case of routers and virtual path identifier/virtual circuit identifier (VPI/VCI) information in case of ATM switches. The soft-state values are cached and can be disposed of if necessary for a new flow or a connection. The active network connections are more efficient and intelligent; the state left behind is in the form of program codes rather than static table entries. The capsules can be used in some joint computation, such as the pruning of multicast trees. What happens is that the capsules can be programmed to meet at a node by arranging for the first capsule to get some state information and then waiting for other capsules to arrive. Then these can get engaged in joint task computation.

2.6.2.3 Program Extensibility

Extensibility is required in case of active nodes so that the capsules can install uniquely named classes of C++ and methods, in case of Java, at nodes for reference or use by other capsules and methods. To speed up the loading process, the nodes could contain a cache of known external methods or classes and be equipped with a scheme that facilitates it to locate and dynamically load methods or classes on demand. This dynamic approach is more flexible with respect to timing and decision. The cost of this dynamic approach is the requirement to ensure safety while code is loaded. The discrete approach ensures that program loading must be completed before usage.

2.6.2.4 Interoperability in Capsule-Based Networks

Traditional packet networks achieve interoperability by standardizing the syntax and semantics of packets through reference network models. For example, all IP routers support the agreed upon IP specifications; although router designs of different vendors may vary, they roughly execute the same programs. The active nodes, in contrast, implement very different programs, i.e., they treat the packets flowing through them very differently. Network interoperability is attained at a higher level of abstraction. This is in contrast to standardizing the computation performed on every packet. The middleware and Java Virtual Machines (JVMs) discussed in Chapter 3 help in achieving this interoperability. For interoperability reasons, all of the active nodes along a capsule's path should be able to evaluate the capsule's contents. One of the ways of achieving this is to use the same programming language to express the programs, and the other approach is to use platform-independent functionality of a language, for example JVM. The latter approach is sometimes difficult to achieve due to the heterogeneity of programming languages and hardware systems.

2.6.2.5 Enabling Active Technologies

These technologies enable safe execution of programs by restricting the set of actions available to mobile programs and the scope of their operands.

2.6.2.6 Source Code

Safe-Tcl is a language that achieves safety through interpretation of a source program and restricted access of its namespace.[21] This property prevents programs from straying beyond the transient execution environment. Additionally, Tcl's character-based representation makes it easy to design programs that create new source fragments. The main disadvantage here is the overhead of source code interpretation, which is compounded by Tcl's encoding of all data. The other disadvantage is the huge size of programs. The programs can be compressed, but readability becomes poor.

2.6.2.7 Intermediate Code

Java achieves mobility through the use of an intermediate instruction set.[21] Java enhances the mobility by loading off some of the responsibility from the interpreter. The instruction set and its approved usage are designed so as to reduce the degree of operand validation that the interpreter must achieve as each instruction is executed. This is made possible by the static inspection of the code before it is first executed, so that many of the checks need to be performed only once, when the program is first loaded.

2.6.2.8 Platform-Dependent (Binary) Code

The platform-dependent binary is directly executed by the underlying hardware so that such program fragments are safely executed. Their use of the instruction set and address space is restricted. Traditionally, operating systems rely on the use of heavyweight processes and hardware-supported address space protection. To improve the performance, lighter-weight approaches have been used. These lightweight mechanisms deploy trustworthy compilers to generate programs that will not cross a restricted execution environment. When a program is loaded, the runtime system verifies that the instruction sequence is compiled by a trusted compiler and has not been changed. The validation of the code can be independently performed by the receiving platform, or the compilers and vendors of programs are authorized to sign for their code. The first approach improves mobility across nodes. The latter approach not only saves the cost of validation, but also allows the compiler to generate the code.

2.6.2.9 Architectural Considerations

Closed network (traditional) architectures separate the upper (end-to-end) layers from the lower layers (hop-by-hop) layers. The network layer bridges these layers and provides interoperability by providing a service that supports the exchange of information between end systems.

In closed networks, the network layer provides interoperability. There is a detailed specification of the syntax and semantics of the IP that is implemented by routers and end systems. Due to the standard syntax and semantics, all the nodes perform the same computation on the packets flowing through them.

Active network nodes, on the other hand, allow many different computations (i.e., execution of different programs) for different groups of users. Hence, network layer functionality in active networks is based on an agreed upon program encoding and computation environment, instead of a standardized syntax and semantics of IP packets. Thus, the level of abstraction of interoperability in active networks is more than in traditional networks.

2.6.3 Programmable Switch Approach

This scheme does not use active packets for service creation. It implements *out-of-band* programming of ATM switches. The examples of this approach are Tempest for ATM at Cambridge[22–25] and Xbind at Columbia University.[26] These architectures provide an open programming environment that facilitates easy creation of network services and mechanisms for efficient resource allocation by using IP as the communication mechanism. In this approach, open programmable interfaces allow applications and middleware to manipulate low-level network resources to construct and manage services. The objectives include open signaling, rapid service creation, and enhanced control over network objects through IEEE Project 1520 and OPENSIG.

2.7 Components and Architecture of Active Networks

2.7.1 Major Components

The major components of an active network node are the node operating system (NodeOS), the execution environments (EEs), and the active applications (AAs). The general organization of these components is shown in Figure 2.3.[27] Each EE exports a programming interface or virtual machine that can be programmed or controlled by directing packets to it. Thus, EE provides an interface through which end-to-end network services can be accessed. The architecture allows for multiple EEs to be present on a single active node.

EEs can be thought of as the programming environments of active networks and are often based on a particular programming language. The EEs accept valid programs and packets, execute them, and send them to other nodes.

Because multiple EEs may run on a single active node, as shown in Figure 2.3, the node runs an operating system (NodeOS) that manages the available resources, i.e., memory regions, central processing unit (CPU) cycles, and link bandwidth, and mediates the demand for these resources

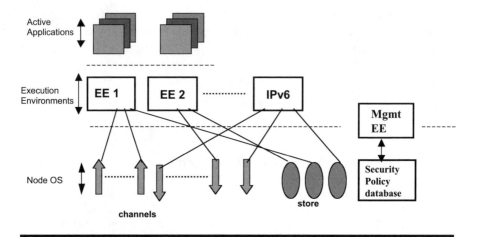

Figure 2.3   Active node components.[27] (© 1998, IEEE)

among EEs running on the node. When an EE requests a service from the NodeOS, the NodeOS presents this information to a security enforcement engine, which verifies its authenticity and checks that the node's security database authorizes the EE or end user to receive the requested service or perform the requested operation. The NodeOS isolates EEs from the details of resource management and from the effects of the behavior of other EEs. Each node has a management execution environment (Mgmt EE), which maintains the security policy database of the node, loads new EEs or updates and configures existing EEs, and supports the instantiation of network management services from remote locations. Any management functions that can be initiated or controlled via packets directed to the management EE must be cryptographically secured and subjected to the polices specified in the security policy database.

2.7.2 Packet Processing in Active Networks

The NodeOS implements communication channels, over which EEs send and receive packets. These channels consist of physical transmission links plus the protocol processing associated with higher-layer protocols (e.g., TCP, User Datagram Protocol (UDP), IP). The general flow of packets through an active node is shown in Figure 2.4.[27] When a packet is received on a physical link, it is first classified based on information in the packet (i.e., headers).

Typically an EE requests creation of a channel for packets matching a certain pattern of headers, e.g., a certain Ethernet type or combination of IP and TCP port numbers. On the output side, EEs transmit packets by

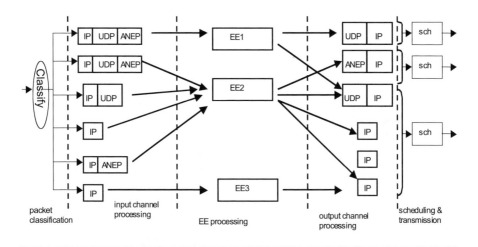

Figure 2.4 Packet flow through an active node.[27] (© 1998, IEEE)

submitting them to output channels, which include protocol processing and packet scheduling. The packets from end users are routed by using the Active Network Encapsulation Protocol (ANEP). The ANEP header contains a Type Identifier field. Packets with a valid ANEP header with a proper type ID can be routed to the appropriate EE. Figure 2.4 shows EE1 receiving ANEP packets encapsulated in UDP datagrams, e.g., those with a particular destination port number. EE2 receives plain UDP datagrams as well as UDP datagrams containing ANEP packets (different port number or ANEP packet type). The EEs also support legacy traffic, i.e., traffic without ANEP packets, simply by setting up the appropriate channels. The ANEP header also provides error handling and node security functions.

An active application (AA) is a program that, when executed by the virtual machine of a particular EE, implements an end-to-end service. Thus, AA implements customized services for end users, using the programming interface supplied by the EE. Details of how the code consisting of AA is loaded into the relevant nodes of the network are determined by the EE; the code may be carried in band with the packet itself or installed out of band. In case there are different EEs at different nodes, an embedded Object Request Broker (ORB) in each node can be used to meet the requirements of interoperability among different AAs from different users.

2.8 Summary

The two approaches that make networks programmable are open signaling (OPENSIG) and active networks. The OPENSIG approach models communication hardware by providing open access to switches and routers. The promoters of OPENSIG approach maintain that by opening up the switches in this way, enhanced control over network objects and development of new and distinct architectures and services, e.g., virtual networking, can be realized. The active networks, on the other hand, propose dynamic deployment of new services on the fly, mainly within the confines of IP networks. Active networking allows intermediate nodes in the network to be programmed dynamically according to user and application needs through active packets (capsules).

The IEEE project 1520 on programmable interfaces for networks has followed the OPENSIG approach in an attempt to standardize programmable interfaces for ATM switches, IP routers, and mobile telecommunication networks.

Exercises

1. Discuss the approaches to introduce programmability into networks?
2. Discuss the three quality of service parameters.
3. How do active and programmable networks tackle the NACK problem?
4. Mention the four types of interfaces in IEEE 1520.
5. Describe the major components and packet processing in active networks. Draw a diagram of packet flow through an active node.

References

1. D.L. Tennenhouse et al., A survey of active network research, *IEEE Commun. Mag.*, 35:1, 80–86, 1997.
2. A.T. Campbell et al., A survey of programmable networks, *ACM SIG-COMM Comp. Commun. Rev.*, 7–23, 1999.
3. Internet Engineering Task Force (www.ietf.org).
4. Stardust.com, *The Need for QoS*, White Paper, July 1999 (www.qosform.com/white-papers/Need_for QoS-v4.pdf).
5. Stardust.com, *QoS Protocols and Architecture*, White Paper, July 8, 1999 (www.qosforum.com; http://www.objectspace.com/products/voyager/).
6. R. Braden, D. Clark, and S. Shenker, *Integrated Services in the Internet Architecture: An Overview*, RFC 1633, Network Working Group, IETF, June 1994 (www.ieft.org).
7. R. Braden, L. Zhang, S. Berson, S. Herzog, and S. Jamin, *Resource ReSerVation Protocol (RSVP): Version 1 Functional Specification*, RFC 2205, Network Working Group, IETF, September 1997 (www.ietf.org).

8. S. Shenker, C. Patridge, and R. Guerin, *Specification of Guaranteed Quality of Service*, RFC 2212, IntServ Working Group, IETF, September 1997.

9. A.J. Wroclawski, *Specification of the Controlled-Load Network Element Service*, RFC 2211, IntServ Working Group, IETF, September 1997.

10. S. Blake, D. Black, M. Carlson, E. Davies, Z. Wang, and W. Weiss, *An Architecture for Differentiated Services*, RFC 2475, Network Working Group, IETF, December 1998 (www.ieft.org).

11. A. Fayaz, A. Shaikh, et al., End-to-end testing of QoS mechanisms, *IEEE Comput.*, 80–86, 2002.

12. S. Floyd and V. Jacobson, Random early detection gateways for congestion avoidance, *IEEE/ACM Trans. Networking*, 1:4, 397–413, 1993.

13. G. Armitage, *Quality of Service in IP Networks Foundations for a Multi-Service Internet*, MTP, 2000.

14. M. Hassan, A. Nayandoro, and M. Atiquzzaman, Internet telephony: services, technical challenges, and products, *IEEE Commun. Mag.*, 38:4, 96–103, 2000.

15. A. Mark and P. Miller, *Voice Over IP*, M&T Books, CA, 2002.

16. S. Keshav, *An Engineering Approach to Computer Networking: ATM Networks, the Internet, and the Telephone Network*, Pearson Education Ltd., Singapore, 1997.

17. D. Wettherall et al., Introducing new Internet services: why and how, *IEEE Network*, 12:3, 12–19, 1998.

18. Z. Fan and A. Mehaoua, Active Networking: A New Paradigm for Next Generation Networks, paper presented at the Second IFIP/IEEE International Conference on Management of Multimedia Networks and Services (MMNS'98), France, November 1998.

19. J. Biswas et al., The IEEE P1520 standards initiative for programmable network interfaces, *IEEE Commun. Mag.*, 36:10, 64–70, 1998.

20. A. Marshall, Dynamic Network Adaptation Techniques for Improving Service Quality, invited paper presented at Networking 2000, Paris, May 2000.

21. D.L. Tennenhouse and D. Wetherall, Towards an active architecture, *Computer Communication Review*, 26:2, 5–18, 1996.

22. J.E. van der Merwe et al., The Tempest: A practical framework for network programmability, *IEEE Network Mag.*, 20–28, 1998.

23. J.E. van der Merwe and I.M. Leslie, Service specific control architectures for ATM, *IEEE J. Selected Areas Commun.*, 16, 424–436, 1998.

24. S. Rooney, The Tempest: a framework for safe, resource-assured programmable networks, *IEEE Commun. Mag.*, 36:10, 42–53, 1998.

25. I.M. Leslie, Shirking networking development time-scales: flexible virtual networks, *Electron. Commun. Eng. J.*, 11:3, 149–154, 1998.

26. A.A. Lazar, Programming telecommunication networks, *IEEE Network*, 11:5, 8–18, 1997.

27. K.L. Calvert et al., Directions in active networks, *IEEE Commun. Mag.*, 36:10, 72–78, 1998.

Chapter 3

Enabling Technologies for Network Programmability

3.1 Introduction

Programmable networks allow dynamic customization of nodes, thus creating new network architectures and services. Software agents and active packets, middleware technology, operating systems, and dynamically reconfigurable hardware have been proposed as the mechanisms that can be employed to achieve this programmability and customization within these networks.[1] This chapter discusses these enabling technologies in detail.

3.2 Enabling Technologies for Network Programmability

3.2.1 Agents

The agent concept, widely proposed and adopted within both the telecommunications and Internet communities, is a key tool in the creation of an open, heterogeneous, and programmable network environment. Agents enhance the autonomy, intelligence, and mobility of software objects and allow them to perform collective and distributed tasks across the network. The term *agent* is used in many different contexts. Thus, there is no single (software) agent definition that exists today. However, agents can be discussed by means of several attributes to distinguish them from ordinary code. It is not essential that all of these attributes be present in a given agent implementation. The single attribute that is commonly agreed upon is *autonomy*. Hence, an agent can be described as a software component that performs a specific task autonomously on behalf of a person or organization.[2] Alternatively, an agent can be regarded as an assistant or helper that performs routine and complicated tasks on a user's behalf. In the context of distributed computing, an agent is an autonomous software component that acts asynchronously on the user's behalf. Agent types can be broadly categorized as static or mobile.[2,3]

Based on these definitions, an agent contains some amount of artificial intelligence, ranging from predefined and fixed rules to self-learning mechanisms.[2] Thus, agents may communicate with the user, system resources, and other agents to complete their task. The agents may move from one system to another to execute a particular task.

3.2.1.1 Agent Technologies

The agent technologies broadly consist of an agent communication language, a distributed processing environment (DPE), and a language for developing agent-based applications.[2] To cooperate and coordinate effectively in any multiagent system, agents are required to communicate with each other. A standard communication language is essential if agents are to communicate with each other. There are two main categories of agent communication: basic communication and communication based on standardized agent communication languages (ACLs). Basic communication is restricted to a finite set of signals. In a multiagent system, these signals have the capability of invoking other agents for a desired action, whereas communication based on ACLs involves the use of a software language for communication between agents.[4] For example, Foundation for Intelligent Physical Agents (FIPA) ACL is mainly based on the Knowledge Query and Manipulation Language (KQML).[4] KQML can be considered to consist of three layers: the content, message, and communication layers. The content layer specifies actual proportions of the message. At the second level, KQML provides a set of performatives that constitutes the message layer (e.g., ask, tell, reply, etc.). A KQML message could contain Prolog code, C code, natural language expressions, or whatever, as long as it can be represented in ASCII. This layer enables agents to share their intentions with others. The communications layer defines the protocol for delivering the message and its encapsulated content. KQML has some shortcomings, the most important of which is its lack of precise semantics.

Languages such as Smalltalk, Java, C++, or C can be used for the construction of agent-based applications. Choice of the language depends on the type of agent functionality. For example, Java or Telescript is suitable for implementing mobile agents, whereas tool command language (TCL) is more appropriate for interface agents.[5]

To provide a standard agent operational environment in relation to underlying communications infrastructure, creation of a DPE is necessary. The creation of a DPE provides physical communication transparency to the agents via Java Remote Method Invocation (RMI) or Common Object Request Broker Architecture (CORBA).[6,7] These provide the supporting infrastructure for a DPE where agents can operate. It may involve the creation of a platform-independent agent environment generated by the execution of, e.g., Java Virtual Machines at hardware nodes.

The majority of current communication system architectures employ the client/server method that requires multiple transactions before a given task can be accomplished. This can increase signaling traffic throughout the network. This problem can rapidly escalate in an open network environment that spans multiple domains. As an alternative solution, mobile agents can migrate the computations or interactions to the remote host by moving the execution there. For example, mobile agents can be delegated to complete specific tasks on their own, provided that a certain set of constraints or rules have been defined for them.[8] They can then be dispatched across the network in the form of mobile program or mobile code that can be recompiled and executed in the remote host.

Agents may be dispatched to the nodes spanning the routes in a network and can be responsible for the maintenance of services through the virtual private networks (VPNs) created. Maintenance of a VPN may involve dynamic reconfiguration and rerouting of connections and renegotiation of quality of service (QoS) targets.[9]

Agents can also be used to implement service level agreements (SLAs). The agent may be a broker or mediator between an end user and a service provider to implement the SLA. In this way, complicated QoS metrics (from the end user's point of view) can be communicated in a simplified manner. Service provider and network provider agents can then negotiate with users' agents to meet the required service.

The serious research activity related to mobile agents started in the early 1990s. A large number of manufacturers were involved in the development of various platforms built on top of different operating systems and based on different programming languages and technologies. However, within the last few years, interpreter-based programming languages like Java have been used for most agent platforms.[2,10,11]

Researchers have also presented several approaches that are associated with the integration of mobile agents and distributed object middleware like CORBA.

To cater for the need of interoperability between agent platforms of different manufacturers, two bodies carry out mobile agent standardization: the Object Management Group (OMG) has defined basic interoperation capabilities between heterogeneous mobile agent platforms in its Mobile Agent System Interoperability Facility (MASIF),[4] whereas the Foundation for Intelligent Physical Agents (FIPA)[4] has focused on the standardization of basic capabilities of intelligent agents. T. Zhang et al. have discussed the integration of these standards and corresponding agent systems.[12]

3.2.1.2 Mobile Agents

According to V.A. Pham and A. Karmouch,[2] mobile agents are self-contained, identifiable computer programs that can move within the network and act on behalf of the user or another entity. To support mobile agents, several fundamental requirements have to be fulfilled by an agent platform. *Continuous support* is required at least for the creation, management, and termination of agents. *Security mechanisms* are essential to protect agent platforms from malicious agents. Finally, a *transport service* is necessary to facilitate agent migration. Application-dependent capabilities are also required to enhance the functionality of the platforms and to support agents in performing specific tasks.

3.2.1.2.1 Applications of Mobile Agents

With the active networks playing a key role in the present and future information systems, it can be easily concluded that mobile agents will be used as part of the solution to implement more adaptable and decentralized network architecture. In this section, the possible applications of mobile agents in network services and network management available in the general literature are presented and evaluated.

Mobile Agents in Telecommunication Services — The current network environment depends on international standards such as the telecommunication management network (TMN) and intelligent network (IN). IN and TMN use the traditional client/server model to provide services via centralized nodes known as service control points (SCPs), which in this case are servers. During a service transaction, the distributed exchanges known as service switching points (SSPs), which are playing the role of clients, ask the SCP for control services so that the SSP can perform actual processing. SCPs and SSPs communicate with each other via a remote procedure call (RPC)-based protocol (the IN Application Protocol (INAP)). IN works by installing IN services, by downloading the necessary IN service components into the IN network elements through specific service management systems (SMSs).

A centralized SCP and the necessary usage of INAP create potential bottlenecks. Therefore, the implementation of services will have to be distributed as close to the customer premises as possible. Mobile agents can enhance the performance by dynamtically downloading the customized *service scripts* (i.e., dynamic migration of service intelligence from the SCP to the SSP), ultimately allowing the provision of *service intelligence on demand,*[2] as shown in Figure 3.1.

Markus Breugst and Thomas Magedanz[3] recognized two general approaches for agent-based service architecture: *smart network* and *smart message.* In the smart network approach, agents are static entities in the network, capable of performing tasks autonomously with other agents and being dynamically configured. This type of agent-based service architecture allows dynamic downloading and exchange of *control scripts*, thus placing intelligence at the network devices. The control scripts can be simple or complex and represent lightweight mobile agents. In the smart message scheme, agents are mobile entities that travel between computers/systems to perform tasks. Agents are received and executed in an agent execution environment (AEE), as shown in Figure 3.2.

The smart message agent can serve as an asynchronous message transporter for its owner (e.g., retrieve e-mail asynchronously and forward it to the current location of the owner) or as a negotiator or broker that requests and sets up all requirements for services (e.g., establishes a real-time connection for media delivery).

With these agent-based approaches, customized and distributed services can be provided at runtime.

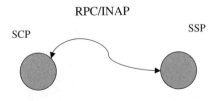

Figure 3.1 Dynamic migration of service intelligence from the SCP to the SSP.

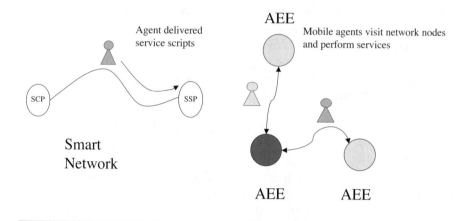

Figure 3.2 The smart network and smart message schemes.

Mobile Agents in Telecommunications Information Networking Architecture (TINA) — TINA, as an evolution from IN and TMN, allows dynamic and transparent distribution of computation objects that are supported by distributed processing environments (DPEs). The TINA Consortium (TINA-C) has worked to accommodate the idea of intelligent and mobile agents as extensions to support TINAs. TINA has identified the following agent dimensions:[2]

- Act on behalf of someone
- Persistent
- Adaptive
- Mobile
- Communicating
- Reasoning
- Environmentally aware
- Socially aware
- Planning
- Negotiating

Mobile Agents in Network Management — The use of mobile agents has been initiated to transform a static-current client/server-based network management into a distributed and decentralized one. This section of the chapter explains the use of mobile agents in network management. Chapter 6 discusses different traditional and active network management schemes.

Traditional network management schemes, such as the Simple Network Management Protocol (SNMP) and Common Management Information Protocol (CMIP), that have been proposed by the International Organization for Standardization (ISO) consist of *management stations* (MSs) and interact with *management agents* running on network nodes. The static agents in these protocols are computational entities or programs responsible for gathering and storing management information local to the node. These agents respond to requests for this information from the MS by using a *management protocol* that specifies the packet format for its communication.

According to Germán Goldszmidt and Yechiam Yemini,[8] this centralized approach in network management seriously hampers its scalability, leading to low performance as the size of the network increases. Recognizing this problem, complex solutions such as notification agents (ISO), *proxy agents* (SNMPv2, Internet Engineering Task Force (IETF)), and *remote monitoring* (RMON, IETF) have been presented to decentralize and relieve the bottleneck around the MS.

In the traditional systems (SNMP, CMIP), the MS does all the computation and sends the results to the devices via fixed-format client/server messages. On the other hand, a decentralized approach was used by Yemini.[8] This approach is called management by delegation (MBD). The MS in MBD sends code and data through agents to be executed at the devices. The executions at the devices would be asynchronous, i.e., not dependent on the MS for code and data. This enables the MS to perform other tasks in parallel to these executions, thus enhancing the performance of the management architecture. The other positive aspect of this architecture is that the codes are not statically bound to any devices; hence, the MS can customize and dynamically enhance the services provided by the agents on any device.[8]

3.2.1.2.2   Mobile Agent Models

There are several mobile agent models that have been developed over the past few years. They have been developed at different universities and companies. Some of these models are discussed here:

- Aglet™ from IBM[13]
- Concordia™ from Horizon Systems Laboratory, Mitsubishi Company[14,15]
- Mole[16]
- Voyager™ from ObjectSpace[17]
- Agent Tcl from Dartmouth College[18,19]
- Agents for Remote Action (ARA) from the University of Kaiserslautern[20,21]

Aglet — An Aglet is defined as a mobile Java object that visits Aglet-enabled nodes in a network. Aglet runs in its own thread of execution or execution environment after arriving at a node, so it is attributed as *autonomous*.[2] The complete Aglet object model consists of additional abstractions such as *context, proxy, message, itinerary,* and *identifier*.[2] An environment is created through these additional abstractions. This provides an opportunity to aglet in which it can carry out its tasks. Aglet uses a simple proxy object to send messages and has a message class to encapsulate message exchange between agents.[13] However, group-oriented communication is not available, and the choice of using a proxy to relay a message may not be a scalable solution. Nevertheless, the Java infrastructure provides the platform independence and mobile code facility.

Concordia — The Concordia agent system is also based on Java. In Concordia, an agent is considered a collection of Java objects. A Concordia agent has been modeled as a Java program that executes in an agent execution environment (AEE) at a given node. These execution environments are formed by a collection of server objects that take care of different functions, such as mobility, persistence, security, communication, and administration of resources. These server objects communicate among themselves and can run in one or several Java Virtual Machines. Once arriving at a node, the Concordia agent accesses standard services available to all Java-based programs, such as database access, file system, and graphics like Aglet.

A Concordia agent has two states: internal and external. The internal states are values of the objects' variables, and the external task states are the states of an itinerary object that would be kept exterior to the agent's code. This itinerary object encapsulates the destination addresses of each Concordia agent and the method that each of these agents would have to execute upon arriving at an execution environment.

Mole — In Mole, the agent is modeled as a hierarchy of Java objects; this implements a closure mechanism without external references except with the host system. The agent is thus a closure over all the objects to which the main agent object contains a reference. Each Mole agent is identified by a unique name provided by the agent system. Also, a Mole agent can only communicate with other agents via defined communication mechanisms, which offer the ability to use different agent programming languages to convert the information transparently when needed.

Voyager — Voyager is a commercial product that introduces itself as an agent-enhanced Object Request Broker (ORB) in Java. Voyager includes several advanced mechanisms that could be utilized to implement MA systems. The Voyager agent model is also based on the concept of a group of Java objects. This system has the facility for moving individual objects and not just the agents. Like Aglet and Concordia, the agent class in Voyager also encapsulates a control model. However, Voyager supports a more extensive set of control mechanisms, such as more flexible instructions on how the agent should terminate itself than the fixed or explicit instructions on termination in Aglet and Concordia.[13,15]

Agent Tcl — The developers of Agent Tcl do not formally specify any particular language to write agents. Instead, agents can be written in any language. Features like mobility are provided by a common service package implemented through an agent on the server. This server offers mobile agent-specific services such as state capture and group communication, as well as more traditional services such as disk access and central processing unit (CPU) cycle. Agent Tcl mobility depends on the closure, which is the Tcl scripts. There are no additional codes to load (i.e., no external references).

Agent for Remote Action (ARA) — The ARA system has a core service layer supporting multiple languages through interpreters. In ARA, the mobile agent is a program that is able to move at its own choice, utilizing various established programming languages. ARA agents execute within an interpreter, interfacing with a core language-independent set of services. The core services include resource management, mobility, and security. ARA agents are executed in parallel threads, while some of the internal ARA core functions can be executed as separate processes for performance reasons.

3.2.1.2.3 Mobile Agent System Requirements

For successful implementation of the mobile agent systems, there are at least nine requirements that should be met:[2]

- Security
- Portability
- Mobility
- Communication
- Resource management
- Resource discovery

- Identification
- Control
- Data management

Security — The security of mobile agents is concerned with the requirements and actions of protecting the AEE from malicious agents, protecting agents from a malicious AEE, and protecting one agent from another. It is also concerned with the protection of the communication between AEEs.

V.A. Pham and A. Karmouch[2] have suggested partitioning the network into one or more domains with a protected computer running an interpreter (i.e., an AEE) that is trusted by all agents in that domain. This approach is similar to the playground approach proposed by D. Malkhi and M. Reiter,[41] where a dedicated machine is reserved for downloading an agent. These special interpreters in various domains trust each other to numerous degrees, depending on the relationships between the domains. All other interpreters are considered unfriendly and dangerous by default and cannot be trusted. Agents in this system require permission to collect audit data and respond to attacks, but at the same time must be controlled so that they will not exceed their right. To be effective, these special interpreters that control the privileges must be well protected. The major problem here is securing the trusted interpreter from a rogue or obscure agent, or a trusted agent that has been assigned to perform malicious tasks. Additionally, ensuring security across the network for all hosts in this type of scheme is difficult.

The traditional security mechanisms for authentication, authorization, and access control use cryptographic techniques. These mechanisms are also used to protect the AEE from malicious agents or security breaches from another AEE, but still there are no satisfactory measures to prevent a hostile AEE from doing damage. The reason is that the internal code of an agent is exposed to an AEE for execution.[2] An unfriendly AEE can attack by not running the agent code correctly, refusing to transfer the agent, tampering with agent code and data, or listening to interagent communication.[2] The current solutions to protect an agent from a hostile AEE proposed by different researchers include cryptographic tracing of execution, duplication of computation and cross-examination, code obfuscation, secure domains using a co-processor, and providing warnings about possible attacks.[10] Unfortunately, these solutions incur high computation costs.

A secure object space and cryptographic techniques provide reasonable solutions to protect agents from each other. Protecting hosts from malicious agents currently has many solutions. The various schemes to protect hosts from malicious agents are access control, various authorizations and authentications using digital signatures and other cryptographic techniques, and proof-carrying code.[9]

Most of the above-mentioned mobile agent systems are implemented in Java; the implementers of these systems depend on the Java security model and provide customized extensions to this model. Aglet has its own implementation of the security manager and uses the sandbox Java applet model. Mole does not yet have any security model other than the basic Java facility. Concordia also has its own security management and implements a secure transport mechanism with the secure socket layer.[15] Voyager works with a security manager plus a customizable socket interface that developers can employ to implement arbitrary socket-level security. The security mechanisms in ARA and Agent Tcl are not yet mature enough.

Portability — Portability deals with heterogeneity of platforms and with porting agent code and data to work on multiple platforms. Before Java, agent systems such as Agent Tcl and ARA depended on an operating system (OS)-specific core and language-specific interpreters for agent code and data.[20] The Java Virtual Machine (JVM) presents a better solution, as platform-independent bytecode can be compiled just in time to deal with heterogeneity of platforms.

Mobility — Most systems mentioned in the previous section use application protocols on top of Transmission Control Protocol (TCP) for transportation of agent codes and states. Because these systems are based on Java, they use Remote Method Invocation (RMI), object serialization, and reflection as the mechanisms for mobility. Usually the agent states and codes are converted into an intermediate format to be transported and restarted at the other end.

Aglet implements its own Agent Transfer Protocol (ATP). This is an application-level protocol that communicates via a TCP socket. Concordia also utilizes TCP sockets and Java object serialization for transport mechanisms. Voyager employs Java object serialization and reflection largely in its transport mechanism. Both ARA and Agent Tcl may utilize a number of protocols to transfer agent code and data. Agent Tcl currently employs a TCP socket, while ARA uses the Simple Mail Management Protocol (SMMP), Hypertext Transfer Protocol (HTTP), and File Transfer Protocol (FTP).

Communication — Systems based on Java mostly support event, message, or RMI-based communication, while ARA supports only client/server-style message exchange at a predefined service point. ARA still does not permit interagent communication. Agent Tcl supports RPC-style communication between agents and a message-passing model via byte streams.[18]

Agent Tcl has introduced an Agent Interface Definition Language (AIDL) to permit agents in different languages to communicate.[18] The AIDL approach is similar to CORBA IDL. This method has the advantage of interface matching communication that facilitates client/server bindings. However, because it is based on RPC, interfaces have to be established before a proper connection is established.

Other agent systems, described above, that are based on Java make use of the homogeneous language environment to bypass the interface incompatibility issue.

Identification — The basic requirement is that the agents must be identified uniquely in the environment in which they operate. Proper identification permits communication control, cooperation, and coordination of agents to take place.[2]

Different agent systems have different mechanisms for identification of agents. Mole uses Domain Name System (DNS) to assign identifiers or names to agents, and Voyager has an alias facility that a programmer can use to refer to an agent. Voyager also supports a naming service that allows linking of directory services to form a large logical directory (DNS style). This facilitates in locating an agent. In contrast, Aglet uses a simple table lookup to associate name string to universal resource locater (URL), while Concordia uses a directory manager to manage naming service for its agents.[15]

Control — Control in agent systems is based on the coordination of mobile agent activities. Furthermore, it provides means through which agents may be created, started, stopped, duplicated, or instructed to self-terminate. Voyager gives more freedom for control. Any method can be called remotely, and it also has an event multicast facility to allow for control on a group basis.

3.2.2 Middleware Technology

Middleware is a class of software technologies designed to help manage the complexity and heterogeneity inherent in distributed systems. It is defined as a layer of software above the operating system but below the application program that provides a common programming abstraction across a distributed system.

Network applications communicate with one another to deliver high-level services, such as controlling a multimedia communication session or managing a set of physical or data resources. The creation of QoS-based communication services and the management of network resources can be conflicting objectives that network designers have to meet. Middleware technology provides a logical communication mechanism among network entities by hiding the underlying implementation details of different hardware and software platforms. There are different types of middleware, e.g., distributed tuples, remote distributed object middleware (CORBA, Distributed Component Object Model (DCOM)), etc.

The Common Object Request Broker Architecture (CORBA) from the Object Management Group (OMG) has been widely considered to be the choice architecture for the next generation of network management.[22] OMG is based on its Object Management Architecture (OMA), and CORBA is one of its key components. The most significant feature of CORBA is its distributed object-oriented (OO) architecture. This extends the OO methodology into the distributed computation environment. Its advantages also include interoperability and independence of platform. The Object Request Broker (ORB) is used as the common communication mechanism for all objects. OMG defines a stub as the client-end proxy and a skeleton as the server-end proxy. Using the ORB, a client can transparently invoke a method on a server object, which can be on the same machine or across the network. From the CORBA 2.0 release onward, every ORB is expected to understand at least the Internet Inter-ORB Protocol (IIOP), which is layered on top of TCP.[23]

Until now, most of the research has been performed using CORBA to integrate incompatible legacy network management systems (e.g., SNMP and CMIP) and the definitions of specific CORBA services (e.g., notification, log).[24] In addition to the basic features of CORBA, the OMG has defined some services that can provide a more useful API.[24] Among them, the naming and life cycle services are basic CORBA services that are used by almost all CORBA applications. There are already some free and commercial versions of CORBA-SNMP and CORBA-CMIP gateways available.

Different active and programmable projects, for example, Spawning Networks[25] and Tempest,[26] use CORBA as the middleware technology. In Spawning Networks, CORBA provides the necessary middleware support for the interaction of all system objects (e.g., routelets) that characterize a new child virtual network architecture. Tempest uses CORBA to provide an interface between different control architectures. In future networks, CORBA middleware can also be used as a communication interface for different agents and for passing them on to the network and physical layers.

3.2.2.1 Object Management Architecture

The OMA consists of an *object model* and a *reference model*. The object model describes how objects distributed across a heterogeneous environment can be classified, while the reference model distinguishes interactions between those objects.

In the OMA object model, an object is an encapsulated entity with a distinct absolute identity whose services can be accessed only through well-defined *interfaces*. Clients send requests to objects to perform services on their behalf. The implementation and location of each object are concealed from the requesting client.

Figure 3.3 depicts the components of the OMA reference model. The Object Request Broker (ORB) component takes care of the communication between clients and objects. Utilizing the ORB components, there are four object interface categories, described below.

3.2.2.1.1   Object Interface Categories

Object Services — These are domain-independent interfaces that are used by many distributed object programs. The domains will include type

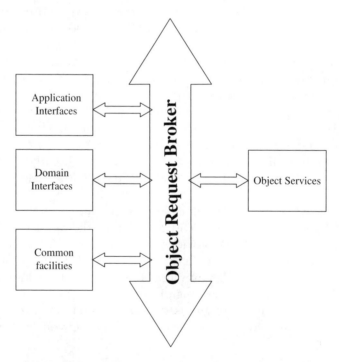

Figure 3.3   The components of the OMA reference model.

of applications, e.g., telecommunications, finance, and medical, running on different platforms. Two examples of object services that fulfill this role are:

Naming service: Permits clients to find objects based on names.

Trading service: Permits clients to find objects based on their properties.

Common Facilities — These are oriented toward end-user applications. An example of such a facility is the Distributed Document Component Facility (DDCF).[27] DDCF permits the presentation and interchange of objects based on a document model, for example, facilitating the linking of a spreadsheet object into a report document.[27]

Domain Interfaces — These interfaces have roles similar to those of object services and common facilities, but are geared toward specific application domains, for example, domain interfaces for product data management (PDM) enablers for the manufacturing domain.[27,28]

Application Interfaces — These are interfaces developed specially for a given application. Because they are application specific, and because the OMG does not develop applications (only specifications), these interfaces are not standardized.

3.2.2.2 The Common Object Request Broker Architecture

When the Open System Interconnect (OSI) model was presented, dealing with heterogeneity in distributed computing systems had not been easy. Particularly, the development of software applications and components that support and make efficient use of heterogeneous networked systems was very difficult. Many programming interfaces exist for developing software for a single homogeneous platform. However, few help deal with the integration of distinct systems in a distributed heterogeneous environment.

To deal with these problems, the Object Management Group (OMG) was created in 1989 to develop, adopt, and promote standards for the development and deployment of applications in distributed heterogeneous environments.

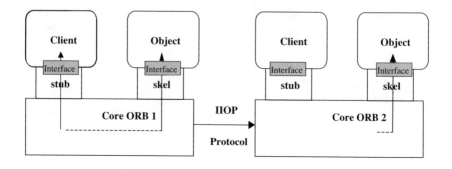

Figure 3.4   The components of CORBA.

One of the first specifications to be adopted by the OMG was CORBA. It gives details of the interfaces and characteristics of the ORB component of the OMA. The main features of CORBA are:

- ORB core
- OMG Interface Definition Language (OMG IDL)
- Stubs and skeletons
- Dynamic invocation and dispatch
- Object adapters
- Inter-ORB protocols

These components are shown in Figure 3.4, which also shows how the components of CORBA relate to one another.

3.2.2.2.1   ORB Core

As mentioned above, the ORB delivers requests to objects and returns any responses to the clients making the requests. The object to which a client wants the ORB to direct a request is called the *target object*. The key feature of the ORB is its transparency in providing client/object communication. The ORB provides transparency for the following:[27]

> **Object location**: The client does not know the location of the target object. It could exist in a different process on another machine across the network, on the same machine but in a different process, or within the same process.[27]
> **Object implementation**: The client does not know how the target object is implemented, that is, in which programming or scripting languages it was developed, nor the operating system (if any) and hardware on which it executes.[27]

Object execution state: When a client makes a request on a target object, the client does not need to know whether that object is currently activated (i.e., in an executing process) and ready to accept requests. The ORB transparently starts the object if necessary before delivering the request to it.

Object communication mechanisms: The client is not aware of the communication mechanisms (e.g., TCP/IP, shared memory, local method call) the ORB uses to deliver the request to the object and return the response to the client.

To make a request, the client specifies the target object through an *object reference*. When a CORBA object is created, an object reference for it is also produced. An object reference always refers to the single object for which it was created as long as that object exists. Object references are both absolute and opaque, so that a client cannot modify them. Only an ORB knows what constitutes an object reference. Object references can have standardized formats, such as those for the OMG standard *Internet Inter-ORB Protocol* and *Distributed Computing Environment Common Inter-ORB Protocol*, or they can have proprietary formats. Clients can obtain object references in several different ways.

Object creation: There are no special client operations for object creation; objects are created by invoking creation requests. A creation request returns an object reference for the newly created object to the client.

Directory service: A client can invoke directory lookup services of some kind to obtain object references. The naming and trader services mentioned above allow clients to obtain object references by name or by properties of the object. Clients store object references and associated information (e.g., names and properties) for existing objects and deliver them upon request.[27]

Convert to string and back: An application can request the ORB to turn an object reference into a string, and this string can be stored into a file or database. Afterwards, the string can be retrieved from persistent storage and turned back into an object reference by the ORB.[27]

In summary, the function of the ORB is to simply provide the communication and activation infrastructure for distributed objects.

3.2.2.2.2   OMG Interface Definition Language

Before a client can make requests on an object, it must know the kinds of operations supported by the object. An object's *interface* identifies the operations and types that the object supports, and thus describes the requests that can be made on the object. Interfaces for objects are defined in the OMG Interface Definition Language (OMG IDL). Interfaces are similar to classes in C++ and interfaces in Java. An example OMG IDL interface definition is[27]

```
// OMG IDL
interface Active
{ Object create();
};
```

This IDL specifies an interface named Active that supports one operation, create. The create operation returns an object reference of type Object. Given an object reference for an object of type Active, a client could invoke it to create a new CORBA object.

An important aspect of OMG IDL is its *language independence.*[27] Because OMG IDL is a declarative language, not a programming language, it enables interfaces to be defined independently from object implementations. This permits objects to be constructed using different programming languages and yet still communicate with one another. Language-independent interfaces are critical within heterogeneous systems, because not all programming languages are supported or available on all platforms.

OMG IDL offers a set of types that are similar to those found in a number of programming languages. It provides basic types such as long, double, and Boolean, constructed types such as *struct* and discriminated union, and template types such as sequence and string. Types are used to specify the parameter types and return types for operations. In the example above, operations are used within interfaces to specify the services provided by those objects that support that particular interface type.[27]

3.2.2.2.3   Stubs and Skeletons

OMG IDL compilers and translators generate client-side *stubs* and server-side *skeletons*. A stub is a procedure that effectively creates and issues requests on behalf of a client, while a skeleton is a mechanism that delivers requests to the CORBA object implementation. Because they are translated directly from OMG IDL specifications, stubs and skeletons are normally interface specific.[27]

Delivery through stubs and skeletons is often called *static invocation.* OMG IDL stubs and skeletons are built directly into the client application and the object implementation. Therefore, they both have complete *a priori* knowledge of the OMG IDL interfaces of the CORBA objects invoked.

The stub works in conjunction with the client ORB to *marshal* the request. That is, the stub helps to convert the request from its representation in the programming language to one suitable for transmission over the connection to the target object. Once the request arrives at the target object, the server ORB and the skeleton work together to *unmarshal* the request (convert it from its transmissible form to a programming language form) and send it to the object. Once the object completes the request, any response is sent back the way it came: through the skeleton and the server ORB, over the connection, and then back through the client ORB and stub, before finally being returned to the client application. Figure 3.4 shows the positions of the stub and skeleton in relation to the client application, the ORB, and the object implementation.

This discussion depicts that stubs and skeletons play important roles in connecting the programming language constructs to the underlying ORB. The stub converts the function call style of its language mapping to the request invocation mechanism of the ORB. The skeleton adapts the request-dispatching system of the ORB to the method expected by the object implementation.[27]

3.2.2.2.4   Dynamic Invocation and Dispatch

In addition to static invocation via stubs and skeletons, CORBA supports two interfaces for dynamic invocation:[27]

- *Dynamic invocation interface* (DII): Provides dynamic client request invocation.
- *Dynamic skeleton interface* (DSI): Supports dynamic dispatch to objects.

The DII and DSI can be considered a *generic stub* and *generic skeleton,* respectively. Each of these interfaces is provided directly by the ORB.

Dynamic Invocation Interface — Through the DII, a client application can invoke requests on any object without having compile-time knowledge of the object's interfaces. It avoids the need to recompile a program to include new static stubs every time a new CORBA object is created. Instead, the program can use DII to invoke requests on any CORBA object at runtime. The DII is also applied to interactive programs such as browsers, which can obtain the values necessary to supply the arguments for the object's operations from the user.

Types of Request Invocation — Following are the types of request invocations through a client:

> **Synchronous invocation**: The client invokes the request and then blocks, waiting for the response until its message is stored in a local buffer at a receiver or actually delivered to the receiver. From the client's perspective, this is effectively the same in behavior as an RPC. This is the most common invocation mode used for CORBA applications because it is also used by static stubs.
>
> **Deferred synchronous invocation**: In this type of invocation, the client invokes the request, continues processing while the request is dispatched, and later collects the response. This is helpful if the client has to invoke a number of independent long-running services. This scheme improves performance by invoking each request serially and blocking for each response; also, all requests can be issued simultaneously, and responses can be collected as they arrive.
>
> **One-way invocation**: The client invokes the request and then continues processing. The client does not wait for the acknowledgment from the server. This type of invocation has reliability problems.
>
> **Dynamic skeleton interface** (DSI): Similar to the DII is the server-side DSI. Just as the DII allows clients to invoke requests without having access to static stubs, the DSI allows servers to be written without requiring skeletons for the objects compiled statically into the program. The main reason for its introduction was to support the implementation of gateways (working on both sides, translating for clients and servers) between ORBs utilizing different communications protocols. Later on, inter-ORB protocols were introduced for ORB interoperation. Nevertheless, the DSI is still a valid choice for a certain class of applications, especially for bridges between ORBs and for applications that serve to bridge CORBA systems to non-CORBA services and implementations.

3.2.2.2.5   Object Adapters

An object adapter or object wrapper is an entity that adapts the interface of another object to the interface required by a caller. In other words, it is an interposed object that transparently allows a caller to invoke requests on an object even though the caller is not aware of the object's true interface.

Object adapters perform the following additional duties:

> **Object registration**: Object adapters supply procedures that facilitate programming language entities to be registered as implementations for CORBA objects. Details of exactly what is registered and how the registration is achieved depend on the programming language.
>
> **Object reference generation**: Object adapters generate object references for CORBA objects. These are absolutely critical in directing requests to objects.
>
> **Server process activation**: If required, object adapters start up server processes in which objects can be activated.
>
> **Object activation**: Object adapters activate objects if they are not already active when requests arrive for them.
>
> **Request demultiplexing**: Object adapters interoperate with the ORB to ensure that requests can be multiplexed over multiple connections without being blocked.
>
> **Registered object requests**: Object adapters send requests to registered objects.

Because CORBA supports diverse object implementation styles, the absence of an object adapter would imply that object methods (implementations) would connect themselves directly to the ORB to receive requests. Having a standard set of just a few object call interfaces would mean that only a small number of object methods (implementations) could ever be supported. Alternatively, standardizing many object call interfaces would add unnecessary complexity to the ORB itself. Therefore, CORBA allows for multiple object adapters. A different object adapter is normally essential for each different programming language. CORBA initially provided Basic Object Adapter, but this failed to perform all object implementations because object adapters tend to be very language specific due to their close proximity to programming language objects.

3.2.2.2.6   Inter-ORB Protocols

Interoperability in CORBA has been a major concern for researchers for some time. Initial commercial ORB products did not have the provision of interoperability. Lack of interoperability was caused by the fact that the CORBA 2.0 specification did not direct any particular data formats or protocols for ORB communications.

CORBA 2.0 introduced a general ORB interoperability architecture that provides for direct ORB-to-ORB interoperability and bridge-based interoperability. Direct interoperability is possible when two ORBs reside in the same *domain*, that is, they understand the same object references and share other information. Bridge-based interoperability is critical when ORBs from separate domains are required to communicate. The function of the bridge is to map ORB-specific information from one ORB domain to the other.

The general ORB interoperability architecture is based on the General Inter-ORB Protocol (GIOP),[27,29] which identifies transfer syntax and a standard set of message formats for ORB interoperation over any connection-oriented transport (e.g., TCP).

The Internet Inter-ORB Protocol (IIOP) identifies how GIOP is developed over TCP/IP. The ORB interoperability architecture also provides for other environment-specific inter-ORB protocols (ESIOPs). ESIOPs allow ORBs to be built for special situations in which certain distributed computing infrastructures are already in use. The first ESIOP, which utilizes the Distributed Computing Environment (DCE),[27] is called the DCE Common Inter-ORB Protocol (DCE-CIOP). It can be used by ORBs in environments where DCE is already installed. This allows the ORB to use existing DCE functions, and allows for easier integration of CORBA and DCE applications.

In addition to standard interoperability protocols, standard object reference formats are also necessary for ORB interoperability. Although object references are transparent to applications, ORBs use the contents of object references to help find out how to direct requests to objects. CORBA specifies a standard object reference format called the Interoperable Object Reference (IOR). An IOR stores information needed to locate and communicate with an object over one or more protocols. For example, an IOR containing IIOP information stores host name and TCP/IP port number information. Most commercially available ORB products support IIOP and IORs and have been tested to ensure interoperability.

3.2.3 Operating System Support for Programmable Networks

An operating system forms the layer between a middleware and the hardware. It basically performs resource management and control of untrusted active applications. This section is based on the discussion of application of the operating system in active networks.

As discussed in Chapter 2, the intermediate (router) node in an active network can be expressed in terms of a model that divides the system into three logical layers:[31] NodeOS, execution environment (EE), and active application layer.

The NodeOS layer abstracts the hardware and provides low-level resource management functionality. An execution environment (EE) is situated above the NodeOS and provides the basic application programmable interface (API) available to the active network programmer. For example, an EE may be a Java Virtual Machine. Conceptually, there may be several EEs running above the NodeOS, each providing a separate programming environment. The third and uppermost layer of the architecture comprises the active applications (AAs), each of which contains code.

3.2.3.1 Janos: A Java-Oriented OS for Active Network Nodes

Janos was presented by P. Tullmann et al.[32] It is designed to prevent separate active applications from interfering with one another and to provide node administrators with strong controls over active applications' resource usage. The Janos architecture was derived from the Defense Advanced Research Projects Agency (DARPA) active node architecture and consists of the following components:[32]

- ANTSR
- Janos Virtual Machine
- Moab (NodeOS Api in C)
- OSKit

The uppermost layer of the architecture comprises the active applications (AAs), each of which contains code. Figure 3.5 shows the architecture of Janos.

Active Applications
ANTSR
Janos Virtual Machine
OS API
OSKit

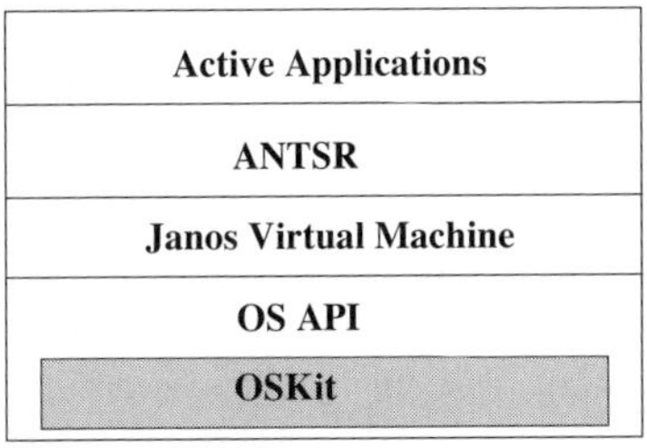

Figure 3.5   The architecture of Janos. (© 2001, IEEE)

3.2.3.1.1   ANTSR

The ANTSR Java runtime is based on ANTS; it provides the interfaces for untrusted, and possibly malicious, AAs to interact with the system.[32] This layer is responsible for hiding essential Janos Virtual Machine (JANOSVM) interfaces and specifying per domain resource boundaries. ANTSR has been developed entirely in Java. ANTSR supports active packet streams, where code is dynamically loaded on demand at runtime when packets for a new stream arrive.[32]

ANTSR differs internally from ANTS in that ANTSR takes advantage of the NodeOS abstractions and the support provided by the JANOSVM. ANTSR consists of new features, including domain-specific threads, separate namespaces, improved accounting over code loading, and a simple administrator's console.[32]

Only authorized users, i.e., node administrators in ANTSR, are allowed to restrict unauthorized or malicious code by hiding interfaces or shutting down a node. The administrator of a Janos node, in addition to initiating the bootstrap process, has the authority to control access to the node, assign resources, and query the node about its state. In Janos, an administrator is provided a control interface to the ANTSR runtime. From there, the administrator can initiate and finish domains, collect statistics about the node, and modify access privileges. Furthermore, simple facilities for maintaining soft state and logging are also provided.

3.2.3.1.2   Janos Virtual Machine (JANOSVM)

The JANOSVM is the most important part of a Janos node. It maps the C-based Moab interfaces into Java and provides ANTSR with support for managing untrusted, potentially malicious, user applications.[32]

The JANOSVM is basically a virtual machine that accepts Java bytecodes and executes them on Moab. The JANOSVM provides access to the underlying NodeOS interfaces through the Janos Java NodeOS bindings, which wrap simple Java classes around the C-based API.

The JANOSVM is based directly on the KaffeOS[32] implementation. KaffeOS is a JVM that provides the ability to isolate applications from each other and to limit their resource consumption. Each process (domain in JVM) in KaffeOS executes as if it were running in its own virtual machine, including separate garbage collection of its own heap. The *domain* is the unit of resource (memory) control. Each domain is associated with a *memory pool* of physical memory pages. A domain contains *thread pools* from which *thread objects* are taken to handle *packets* dispatched out of *input channels*. Packets are delivered on *output channels*.[32]

As far as safety in the JANOSVM is concerned, it supports a type-safe environment through Java; a Java-level runtime (ANTSR in this case) is required to present AAs with restricted access to the NodeOS abstractions and provide services such as protocol loading.[32]

Strict Separation of Flows — The JANOSVM implements a strict separation of domains or processes. Each domain executes in its own namespace and in its own heap. Namespace separation is achieved by a class loader, the standard Java namespace control mechanism. The only shared objects allowed between domains are packet buffers.

The JANOSVM offers each domain with its own heap and a separate garbage collection thread for cleaning that heap. In addition to separating the memory usage of each domain, separate heaps implicitly limit the garbage collection costs sustained by each domain. Internally, the JANOSVM groups similar-size objects together on each memory page, which can cause fragmentation of memory. This procedure of maintaining memory ownership on a per page basis greatly simplifies memory recovery upon domain termination, as the JANOSVM can move whole pages into the free page list.

The strongly implemented separation between domains settles the difficult problem of fine-grained sharing in a type-safe system that supports termination. The JANOSVM itself enforces per domain memory limits. When a domain is terminated, the JANOSVM has to return the same number of pages as it acquired when the domain was created.

Although total segregation of domains makes resource management and domain termination simpler, it also makes the system less flexible. So, the JANOSVM provides for restricted interdomain communication.

3.2.3.1.3   OSKit

The OSKit provides full support in the form of device drivers, boot loaders, remote debugging, and thread implementation to upper layers. The OSKit has been written in C.

3.2.3.2 *Bowman: Operating System for Active Nodes*

Bowman is a node operating system.[33] It is a software platform for the implementation of execution environments within active nodes. Bowman was specifically designed and implemented to serve as a platform for the composable active network elements (CANEs)[33] (explained in Chapter 4) execution environments (EEs). However, it can also be implemented as a general platform over which other EEs may also be implemented. Bowman provides the following additional support to EEs:

- Packets are classified by an efficient and flexible packet classification mechanism. Computation for a flow occurs in its own set of compute threads. Hence, the internal packet-forwarding path of Bowman is inherently multithreaded.

■ Bowman provides cut-through channels for packets that do not require per flow processing (these are the paths that do not incur the overheads of multithreaded processing).

■ Bowman provides support for multiple simultaneous abstract topologies, that is, overlay network abstractions, that can be used to implement virtual private networks.

The major components of a Bowman node are communication protocol implementations, data structures, code-fetch mechanisms, per flow processing, and output queuing mechanisms, which are loaded dynamically at runtime.

Channels in Bowman support sending and receiving packets via an extensible set of protocols. Bowman exports a set of functions enabling EEs to create, destroy, query, and communicate over channels that implement traditional protocols (TCP, User Datagram Protocol (UDP), etc.) in various configurations. Bowman also supports data compression and error correction.

The Bowman implementation is layered on top of a host operating system that provides the lower-level services. These low-level services include memory management, thread creation, scheduling, and synchronization primitives. Hence, Bowman operates by providing a uniform active node OS interface over different hardware and operating systems.

The code for an EE can be loaded as an extension to Bowman and typically consists of control code for the EE and a set of routines that implement the user–network interface supported by the EE. When a packet arrives at a Bowman-enabled node, it is classified to decide to which abstract topology it belongs, and the packets that belong to no topology are discarded. If a node wants to send a packet to a neighboring node on a particular abstract topology, the packet is transmitted on a channel that belongs to the topology and has the required node as the other end position. These abstract topologies do not provide any resource guarantees at active nodes or isolation of data between topologies.

Within the underlying Bowman system there are at least three active processing threads. These are a packet input processing thread, a packet output processing thread, and a system timer thread.

Packets received on a physical interface are classified, and a set of channels is identified on which received packets should be processed. With this step, incoming packets are demultiplexed to specific channels where they undergo processing.

In the other instance, packets may be enqueued directly on an output queue without undergoing any further processing. This type of processing is performed through cut-through paths through the Bowman node.

The Bowman design is highly modular and dynamically reconfigurable at runtime. Its design and implementation are decoupled from any particular networking or routing protocol.

3.2.4 Dynamically Reconfigurable Hardware

Programmability and reconfigurability have been used extensively in the communications industry. There are now a number of systems that exploit reconfiguration to adapt their system (including hardware) resources to meet changing requirements.[1] Reconfigurability has been incorporated into products to overcome issues resulting from a lack of standards or the slowness of the standardization process. However, although static reconfiguration of a system's hardware resources is now successfully employed, the concept of dynamically reconfiguring those resources is a popular area for further research. Dynamic reconfiguration implies that the hardware may be reconfigured during runtime. This can be achieved using dynamic reconfiguration logic (DRL) technology such as may be found in the more recent field-programmable gate array (FPGA) families, for example, the Virtex series family by Xilinx,[35] which can be reconfigured within microseconds and permit partial reconfiguration of some part of the device while another part is still operating. This permits the concept of changing or reconfiguring the hardware in real-time as the data is being processed.[36] Runtime reconfigurable FPGAs can be used in active and programmable networks for the dynamic reconfiguration of scheduling mechanisms. FPGAs can be programmed to reconfigure the weights of a scheduler on the fly without stopping it.

3.2.4.1 Applications of FPGAs in Active Networks

The main feature that distinguishes active networks from existing static networks is that the routers in the active networks can be dynamically reprogrammed by third parties, whereas the routers in existing network do not support this functionality. Most of the projects regarding the application of FPGAs to active networks are related to the programmability of routers.

The current routers are based on the best-effort model and do not support high levels of QoS according to varying networks' conditions in terms of traffic, congestion, and other bandwidth requirements. But the commercial use of the Internet makes it essential to change this model and requires introduction of service differentiation among customers. From router perspective, the main mechanisms needed to provide differentiation of services are based on packet classification, buffer management, and scheduling.

The existing routers perform their packet processing operations in custom silicon or application-specific integrated circuits (ASICs).[37] These circuits are optimized to route, filter, queue, and process Internet datagrams in hardware.[37] ASIC chips have very good performance, but these do not support any reprogramability features at the hardware level. Research on active networks has enabled the dynamic downloading of packet processing software. This extraordinary functionality has allowed the functions implemented on the routers to dynamically evolve to handle new protocols and capabilities.

Field-programmable gate arrays (FPGAs) have proved to be an excellent choice technology to develop networks with this functionality. They are optimized to the performance advantage of ASICs and custom silicon, as they can implement parallel logic functions in hardware. They share the dynamic nature of the microprocessors and network processors in that they can be dynamically reconfigured. The main advantage of using reprogrammable logic for networks comes from the fact that complex algorithms for processing packets can evolve by reprogramming the actual hardware on the chip.

When any user needs any additional service from the network, he or she should be able to upload his or her programs to the proposed router without harming the network. The routers can be manufactured in such a way that they can be reconfigured by just altering the logic-level circuits. The few examples are field-programmable port extender (FPX), programmable protocol processing pipeline (P4), flexible high-performance platform (FHiPP), active network processing element (ANPE), and PLATO.

3.2.4.2 Field-Programmable Port Extender (FPX)

3.2.4.2.1 Background Information

The FPX platform was designed and developed by Washington University at St. Louis. It is a part of the Washington University Giga-bit Switch (WUGS).[38] It has been used to perform Internet route lookups, tunnel IPv6 packets, compress, encrypt, and buffer.

3.2.4.2.2   Components of FPX

The FPX is a networking platform that processes packets in reprogrammable hardware. The platform allows modular hardware components to be dynamically loaded into an FPGA device over a network.

The FPX is comprised of two FPGAs: the network interface device (NID) and the reprogrammable application device (RAD). The FPX implements all logic using these two FPGA devices.

Network Interface Device (NID) — The NID controls how packet flows are routed to and from the modules of RAD. The RAD consists of virtual circuit (VC) lookup tables connected to each port to selectively route flows and the control cell processor (CCP). The CCP is concerned with the processing of control cells that are transmitted and received over the network. The NID routes packets to the two modules on the RAD, the network interface to the switch and the network interface to the line card, using a four-port switching core. Each of the NID's four-port interfaces provides a small amount of buffering for short-term congestion of traffic.

Reprogrammable Application Device (RAD) — The re-programmable application device (RAD) is a field-programmable gate array (FPGA).[38] The RAD provides logic and memory resources for network data processing. The entity performing the network data processing is known as a module. The NID writes data into the RAD synchronous RAM (SRAM) using the data and the sequence number of the incoming cell. One very important aspect of FPX is that during the partial reprogramming, the RAD modules can process other packets as well. This is due to the fact that there are commands that only program a portion of the logic on the RAD. The frames of reconfiguration data can be written to the RAD's reprogramming port through NID. Hence, it is possible for the other modules on the RAD to continue processing packets in parallel during the partial reconfiguration. The modules are the intelligent part of the FPX. This is where the processing of packets is performed. The modules are connected to the off-chip memory modules, that is, SRAM and synchronous dynamic RAM (SDRAM). The SRAM is employed for applications that need to implement table lookup operations, e.g., for routing, and the SDRAM is typically utilized for applications that need only transfer bursts of data and can tolerate higher memory latency.[38]

3.2.4.3   P4: Programmable Protocol Processing Pipeline

3.2.4.3.1   Background Information

This P4 architecture is a part of the Switchware project of the Department of Computer and Information Science at the University of Pennsylvania and Bellcore.[39] The aim of the P4 project is the idea that with an intelligent combination of dynamically reconfigurable hardware and software, better performance in terms of QoS and security can be achieved. The Programmable Protocol Processing Pipeline (P4) project[39] has implemented a platform consisting of a set of ALTERA FLEX8000 FPGAs acting as processing elements (PEs) and a switching array selecting which devices are engaged in processing. The PEs are arranged in a pipeline exchanging data through first in first out (FIFO) buffers, and FPGAs can be moved (connected) in and out of the pipeline to be reconfigured. It has been used to implement forward error correction (FEC) protocol processing over a single OC-3 Asynchronous Transfer Mode (ATM) link.

3.2.4.3.2   Basic Architecture of P4

FPGAs in P4 devices are called processing elements (PEs). If some protocol processing elements are necessary, the hardware implementation (i.e., FPGA configuration) will be downloaded into the FPGA device.[39]

P4 comprises a set of RAM-based FPGA devices in a pipeline and a switching array that interconnects FPGAs. PEs form a pipeline chain through the switching array. Each of the PEs (FPGAs) implements one function in the pipeline chain. A PE element reads the data from its attached FIFO buffer (FB), performs its processing, and writes the processed data into the buffer associated with the next PE in the pipeline chain.

The main components of P4 follow.

Input Interface (IIF) — The IIF receives the input stream cells of the processed virtual circuit and stores them in the first PE FIFO buffer of the pipeline. In this way, the IIF selects the virtual circuit to be activated, assuming that the virtual path identifier/virtual circuit identifier (VPI/VCI) corresponds to a single application. The headers of cells (to be boosted) are stripped off in IIF and are directly passed to the output interface (OIF). The payload of the packet is stored in the FIFO buffer of the PE. The main benefit of stripping off the payload from the header is that the extra data (here the header) will not have to be processed in PEs unnecessarily. This will increase the processing capability of the PEs because header reading each time takes away processing power of the PEs.

Output Interface (OIF) — The output interface (OIF) reads an acti-vated cell stream of the VC from its FIFO buffer or from bypass FIFO if there are any cells.

Switching Array — The switching array increases the flexibility of P4. It works on the pipeline ordering.

Processing Element (PE) — PE consists of a FIFO buffer (FB) and an FPGA. The PE reads the data from its FIFO buffer, processes the data, and writes it to the FIFO buffer associated with the next PE in the chain.

Controller — The controller is responsible for the overall control of the processing elements. The controller selects the appropriate set of config-urations and performs the downloads for FPGA devices on demand. During the download by the controller, the FPGA remains in the passive configuration mode. The controller creates the chain with processing elements ordered together using the switching array.[39]

In P4, the header of the packets to be processed is taken off in input interface (IIF). Only the payload is delivered to the processing element (PE). Thus, PE caters to more processing of the data. Eventually, the FPX processes the packet as a whole, so there is no separation of payload from the header to save memory space.

The FPX can perform full processing of packet payloads at rates of more than 2.4 gigabits per second, while the P4 can process data at the rate of 155 megabits per second.[39] Thus, FPX has a better throughput than P4.[39]

The FPX executes its main part of packet processing in RAD modules. On other hand, the processing element and the controller perform the vital part of packet processing in P4. In FPX it is possible to implement multiple modules in RAD to increase throughput of a router. This is very important for hardware flexibility of the router. But the P4 does not have such flexibility.

3.2.4.4   PLATO Reconfigurable Platform for ATM Networks

3.2.4.4.1   Architecture of PLATO

PLATO is a reconfigurable platform for ATM networks.[40] The PLATO platform has a large FPGA (versions with Xilinx Virtex XCV 100 and ALTERA 20K400) that, in addition to the clock generation circuit, program-ming ports, etc., has four main ports:[40]

■ A UTOPIA level 2 port provides the physical connection to daughterboards with copper or optical fiber outputs. The daughterboard consists of ATM framing circuitry.
■ A 256-MB 133-MHz SDRAM port for buffer space.
■ A peripheral component interconnect (PCI) bus port for communication with the host.

The ATM switch is partitioned into three parts: input header processor, multiplexing and application interface (which interfaces with the hardware that makes the network active), and output header processor.

When the UTOPIA level 2 interface of the switch receives an ATM cell, it puts it in a FIFO (there is one FIFO per incoming link). The switch begins to process the cell by disassembling it and loading the respective registers with all the fields of the header (VPI, VCI, payload type (PT), cell loss priority (CLP), and header error correction (HEC)). When the VPI and VCI of a cell have been received, the ATM switch assesses the output VPI, VCI, and the physical link according to an internal lookup table. The ATM switch consists of four internal lookup tables, one per incoming link. Each lookup table is a memory with 256 cell positions each. This way, the switch can handle cells in all four input ports simultaneously. As the header is processed, data from the payload of the cell continues to be received from the UTOPIA level 2 interfaces in parallel for six system cycles. Until the necessary output link is available for data transfer, all the data is stored in an input FIFO buffer.

On the multiplexing and application interface, the control circuitry checks if there are any collisions and allows or rejects cells. This part also collects all the signals from every input and output port of the switch and performs the actual switching of cells. The scheduling of the cells is based on priority, and the priority of a cell to get output permission is based on the time of its arrival. If two or more cells are received on the same cycle from different input ports and they need the same output port for transmission, then the cell that came from the lowest-numbered input port passes and the others are rejected. Applications on the PLATO architecture can have access to 256 MB of Synchronous Dynamic RAM (SDRAM) and can process a large number of cells, or even keep multiple copies of the cells.

The third part of the switch is the output header processor. When a cell gets permission to pass to an outgoing link, the output processor begins to reassemble the header of the cell. It reads the 48-bit bus, loads the respective output registers, and then begins transmission to the UTOPIA level 2 interface.[40]

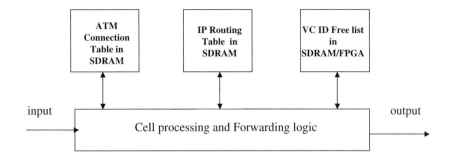

Figure 3.6   The functional diagram of a WIPoATM filter.[40] (© 2001, IEEE)

3.2.4.4.2   Application of PLATO to Wormhole IP over ATM (WIPoATM) Routing Filter

WIPoATM is a routing filter that is situated at the entry points of an ATM network exclusively for the support of IP traffic over ATM. Figure 3.6 shows the functional diagram of a WIPoATM filter. When the first cell of an IP packet arrives, WIPoATM performs the necessary IP routing table lookup using a fast IP routing lookup[40] implemented in hardware and assigns a (preestablished) VP/VC for the transmission of this IP packet. Subsequent cells of this packet will utilize the chosen VP/VC. The process is repeated for the next IP packet, choosing a possibly different VC in a VP from the VC ID free list. If no such unused VC currently exists, then the cells of this packet must be buffered in expectation of some VC getting freed; otherwise, the cell can be dropped.

The ATM connection table and the IP routing table can be stored in the SDRAM modules, while the VC ID list can be kept either in memory inside the processing FPGA or in the SDRAM. While PLATO offers up to four bidirectional 155-Mbps links for input and output, WIPoATM requires two bidirectional links (as it is a bidirectional filter). The remaining two ports can be utilized to implement a second WIPoATM filter on the same prototype, or to double the bandwidth of a single WIPoATM routing filter. The amount of logic offered by PLATO is more than enough for the implementation of two such filters. By managing dynamically a number of VP/VCs, WIPoATM can avoid the (high) latency of establishing connections using ATM signaling.

3.3   Summary

This chapter has described in detail the enabling technologies for active and programmable networks. Software agents facilitate creation of open, heterogeneous, and programmable network environments. Agents play a key role in network management operations and other network services. There are several mobile agent models that have been developed over the last few years. Agent security is a popular area of research these days.

Middleware is a class of software technologies designed to help manage the complexity and heterogeneity inherent in distributed systems. It has two major components: OMG and CORBA. The OMG describes how objects distributed across a heterogeneous environment can be classified. It also distinguishes interactions between those objects. On the other hand, CORBA provides communication and activation infrastructure for these distributed objects by using IDL and inter-ORB protocols.

Janos and Bowman are software platforms for the implementation of execution environments within active nodes. The JANOSVM is basically a virtual machine that accepts Java bytecodes and executes them on Moab. The JANOSVM provides access to the underlying NodeOS interfaces through the NodeOS bindings. Packets in Bowman are classified by an efficient and flexible packet classification mechanism. Hence, the internal packet-forwarding path of Bowman is inherently multithreaded. The major parts of a Bowman node are communication protocol implementations, data structures, code-fetch mechanisms, per flow processing, and output queuing mechanisms.

Runtime reconfigurable FPGAs can be used in active and programmable networks for the dynamic reconfiguration of different functions in routers. The main benefit of using FPGAs for networks comes from the fact that complex algorithms for processing packets can evolve by reprogramming the actual hardware on the chip. FPGA-related projects like field-programmable port extender (FPX), programmable protocol processing pipeline (P4), flexible high-performance platform (FHiPP), active network processing element (ANPE), and PLATO enable reconfiguration of different elements of routers and ATM switches dynamically at runtime.

Exercises

1. Discuss the advantages and disadvantages of mobile agents. How are security issues resolved in applications deploying mobile agents?
2. Discuss the main features of CORBA. What is the significance of IDL in CORBA?
3. Describe the programmable features of Janos and Bowman. Give examples where necessary. What are the security features of Janos?
4. What is the major drawback of P4?

References

1. A. Marshall, Dynamic Network Adaptation Techniques for Improving Service Quality, invited paper presented at Networking 2000, Paris, May 2000.
2. V.A. Pham and A. Karmouch, Mobile software agents: an overview, *IEEE Commun. Mag.*, 36:7, 26–37, 1998.
3. M. Breugst and T. Magedanz, Mobile agents: enabling technology for active intelligent network implementation, *IEEE Network*, 12:3, 53–60, 1998.
4. A. Hayzelden and J. Bigham, Eds., *Software Agents for Future Communication Systems*, Springer-Verlag, Berlin, 1999.
5. H.S. Nwana and M. Wooldridge, Software agent technologies, *BT Technol. J.*, 14, 68–78, 1996.
6. R.H. Glitho, E. Olougouna, and S. Pierre, Mobile agents and their use for information retrieval: a brief overview and an elaborate case study, *IEEE Network*, 16:1, 34–41, 2002.
7. A. Liotta, G. Pavlou, and G. Knight, Exploiting agent mobility for large-scale network monitoring, *IEEE Network*, 16:3, 7–15, 2002.
8. G. Goldszmidt and Y. Yemini, Delegated agents for network management, *IEEE Commun. Mag.*, 36:3, 66–70, 1998.
9. S. Papavassiliou, A. Puliafito, O. Tomarchio, and J. Ye, Mobile agent-based approach for efficient network management and resource allocation: framework and applications, *IEEE J. Selected Areas Commun.*, 20:4, 858–872, 2002.
10. C.-L. Hu and W.-S.E. Chen, A Mobile Agent-Based Active Network Architecture, paper presented at the 7th International Conference on Parallel and Distributed Systems (ICPADS'00), IEEE, 2000, pp. 445–452.
11. F. Baschieri, P. Bellavista, and A. Corradi, Mobile Agents for QoS Tailoring, Control and Adaptation over the Internet: The ubiQoS Video on Demand Service, paper presented at the Proceedings of the 2002 Symposium on Applications and the Internet (SAINT'02), IEEE, 2002, pp. 107–118.
12. T. Zhang, T. Magedanz, and S. Covaci, Mobile Agents vs. Intelligent Agents: Interoperability and Integration Issues, paper presented at the 4th International Symposium on Interworking, Ottawa, Canada, July 6–10, 1998.
13. http://www.trl.ibm.com/aglets.
14. D. Wong et al., Concordia: an infrastructure for collaborating mobile agents, in *Mobile Agents*, Lecture Notes in Computer Science Series 1219, K. Rothermel and R. Popescu-Zeletin, Eds., Springer, Berlin, 1997, pp. 86–97.
15. http://www.meitca.com/HSL/Projects/Concordia.
16. http://www.informatik.uni-stuttgart.de/ipvr/vs/projekte/mole.html.
17. http://www.objectspace.com/voyager.
18. R.S. Gray, Agent Tcl: a transportable agent system, in *Proc. CIKM Wksp. Intelligent Info. Agents*, J. Mayfield and T. Finnin, Eds., 1995.
19. http://www.cs.dartmouth.edu/~agent.

20. H. Peine and T. Stolpmann, The architecture of the ARA platform for mobile agents, in *Mobile Agents*, Lecture Notes in Computer Science Series 1219, K. Rothermel and R. Popescu-Zeletin, Eds., Springer, Berlin, 1997, pp. 50–61.

21. http://www.uni-kl.de/Ag-Nehmer/Projekte/Ara/index_e.html.

22. P. Haggerty and K. Seetharaman, The benefits of CORBA-based network management, *Commun. ACM*, 41, 73–79, 1998.

23. J.P. Redlich et al., Distributed object technology for networking, *IEEE Commun. Mag.*, 36:10, 100–111, 1998.

24. Q. Gu and A. Marshall, Using CORBA's transaction services to enhance QoS management in programmable networks, in *Proceedings of ICT'2002*, Beijing, June 2002, pp. 52–56.

25. A.T. Campbell et al., Spawning networks, *IEEE Network*, 13:4, 16–29, 1999.

26. Leslie et al., The Tempest: a practical framework for network programmability, *IEEE Network Mag.*, 12:3, 20–28, 1998.

27. S. Vinoski, CORBA: integrating diverse applications within distributed heterogeneous environments, *IEEE Commun. Mag.*, 35:2, 46–55, 1997.

28. S. Maffeis and D.C. Schmidt, Constructing reliable distributed communication systems with CORBA, *IEEE Commun. Mag.*, 35:2, 56–60, 1997.

29. M.C. Chan and A.A. Lazar, Designing a CORBA-based high performance open programmable signaling system for ATM switching platforms, *IEEE J. Selected Areas Commun.*, 17:9, 1537–1548, 1999.

30. A. Hughes, Middleware for managing a large, heterogeneous programmable network, *BT Technol. J.*, 20, 117–126, 2002.

31. L. Peterson, Y. Gottlieb, M. Hibler, P. Tullmann, J. Lepreau, S. Schwab, H. Dandekar, A. Purtell, and J. Hartman, An OS interface for active routers, *IEEE J. Selected Areas Commun.*, 19:3, 473–487, 2001.

32. P. Tullmann, M. Hibler, and J. Lepreau, Janos: A Java-oriented OS for active network nodes, *IEEE J. Selected Areas Commun.*, 19:3, 501–510, 2001.

33. S. Merugu, S. Bhattacharjee, E. Zegura, and K. Calvert, Bowman: A NodeOS for Active Networks, paper presented at IEEE INFOCOM, 2000, Vol. 3, pp. 1127–1136.

34. ITU-T (www.itu.int).

35. Xilinx, VIRTEX VCX100 Field Programmable Gate Arrays, Advance Product Specification, Version 1.50, May 1999.

36. J. Villasenor and W.H. Mangione-Smith, Configurable computing, *Sci. Am.*, 276:6, 54–59, 1997.

37. M. Kabir, *Programmable Routers for Active Networks*, ICE CE 2002, Dhaka, Bangladesh, Dec. 2002.

38. D.E. Taylor, J.W. Lockwood, and N. Naufel, RAD Module Infrastructure of the Field-Programmable PorteXtender (FPX), Version 2.0, cs. seas.wustl.edu/ techreportfiles/wucs-01-16.pdf.

39. I. Hadjic and J.M. Smith, *P4: A Platform for FPGA Implementation of Protocol Boosters*, Distributed Systems Laboratory, University of Pennsylvania, Philadelphia, 1997.

40. A. Dollas, D. Pnevmatikatos, N. Aslanides, S. Kavvadias, E. Sotiriades, S. Zogopoulos, K. Papademetriou, N. Chrysos, and K. Harteros, Architecture and Applications of PLATO, a Reconfigurable Active Network Platform, paper presented at the Proceedings of the 9th Annual IEEE Symposium on Field-Programmable Custom Computing Machines (FCCM'01), 2001, pp. 101–110.

41. D. Malkhi and K. Reiter, Secure execution of Java applets using a remote playground, *IEEE Trans. Software Eng.*, 26:12, 1197–1209, 2000.

Chapter 4

Active and Programmable Network Paradigms and Protoypes

4.1 Introduction

Active and programmable networks are an innovation to network architectures in which customized programs are executed within the nodes of the network. This approach accelerates the pace of flexibility and quality of service (QoS) provisioning by decoupling services from the underlying infrastructure.[1,2] Active and programmable networks have recently generated great interest because of these properties. The results of research on active and programmable networks are likely to have an extensive effect on customers, service providers, network providers, and equipment vendors from wired and wireless network domains.[3,4]

The introduction of new services is a difficult task. This requires major developments in the fields of programming languages and designing principles. The realization of these services is dependent on the enabling technologies described in Chapter 3. Before these ideas are put into action, there is a need to understand the limitations of existing networks, such as their strong dependence on fixed protocol stacks. There are numerous research groups who have presented different active and programmable network paradigms.[5,6] These paradigms introduce programmablility into network infrastructures at different levels. These network models have resulted in the controversy regarding performance and security concerns due to insecure code traversing in a network.

A novel QoS architecture and framework for dynamic adaptation of networks using programmable and active networking paradigms, coupled with the use of agent technologies, has been presented in this chapter.[7–9] This framework considers future *open networking* scenarios, where agents can be used to customize network services on behalf of an end user (or acting on behalf of a collection of end users). Such open networks will be heterogeneous environments consisting of more than one network operator. In this environment, agents will act on behalf of users or third-party service providers, to obtain the best service for their clients. The renegotiation of network services and reconfiguration of router resources such as packet scheduling through hierarchical programming interfaces will introduce a much greater flexibility into the network. This framework is compatible with the emerging standard for open systems, the Institute of Electrical and Electronics Engineers' (IEEE) P1520 reference model.[10] The approach also allows for gradual introduction into today's Internet Protocol (IP) network and offers the potential for a smooth migration from static to programmable networks. This framework uses an efficient scheduling algorithm called *active scheduling*, which is described in Chapter 5.

4.2 Types of Active and Programmable Networks

Several research groups are actively involved in designing and developing programmable network prototypes.[5,6] Each project differs from other projects in terms of architectural domain, level, and application of programmability, as shown in Table 4.1. In Table 4.1, the architectural domain gives the targeted architecture or application domain, e.g., signaling management, transport, etc.[5,6] The programmability domain indicates the area where programmability is introduced. The level of programmability gives the method, granularity, and timescale over which new services are introduced into the network infrastructure. The level of programmability varies for different projects. For example, Active Network Transfer System (ANTS)[11] employs a highly dynamic approach of programmability using capsules carrying data for the reconfiguration of services, whereas Xbind[12] uses a remote procedure call (RPC) model that is simple and static in nature. These two approaches represent two extremes of network programmability.

The programmable network prototypes in Table 4.1 have different programmability domains, e.g., Internet protocols, control plane, flows, etc.; however, none of these projects explicitly propose programmability of scheduling mechanisms except the programmable framework.

Switchlets in Tempest[13–16] introduce programmability in the control plane of Asynchronous Transfer Mode (ATM) switches. Routelets in spawning networks[17,18] aim for the creation and deployment of new network architectures by using priority queuing mechanisms and packet classification in parent routelets. Darwin[19] uses hierarchical fair service curves (H-FSCs) as the scheduling mechanism to introduce priority among services. VAN[20] introduces the management framework for active networks that allows customers to deploy and manage their own active services through the virtual private networks. ANN[21] proposes a high-performance Active Network Node that supports gigabits of traffic and provides automatic and rapid protocol deployment, and application-specific data processing and forwarding. It uses a gigabit ATM switch fabric with deficit round-robin (DRR) as the scheduling mechanism for applications.

4.2.1 The Binding Model

The binding model is a conceptual framework for creation, placement, and management of real-time multimedia services over ATM networks.[31] The ultimate aim of the model is creation of multimedia-distributed services with end-to-end QoS guarantees across heterogeneous networking platforms. The model provides an open programming environment for the creation of distributed services. The binding architecture supports application programming interfaces (APIs) for resource access and management. The service specification and creation process is carried out at the application level by a high-level language. These services need support from four dimensions for (1) name and resource mapping into physical resources, (2) resource management and reservation for QoS guarantees, (3) a media stream transport, and (4) service management functionality for monitoring and control.

4.2.1.1 The Binding Architecture

The binding architecture consists of an organized set of interfaces called the binding interface bases (BIBs). Algorithms run on top of these BIBs. Binding algorithms facilitate the service creation process by interconnecting (binding) network resources. QoS is explicitly provided through a set of abstractions that represent the available resources.

Xbind is an example of a binding architecture that has been implemented at Columbia University by A.A. Lazar.[31–32] This is basically a middleware toolkit for creating scalable multimedia distributed services

Table 4.1 Active and Programmable Network Prototypes and Projects

Projects	Architectural Domain	Networking Technology	Programmability Domain	Programmability Level
Switchlets[13–16]	Enabling distinct control architectures	ATM/IP	Control plane	Semidynamic, open signaling
Routelets[17,18]	Spawning VPNs	IP	Routers in spawning networks	Dynamic
Darwin[19]	Resource management	IP	Flows	Semidynamic
Supranet[22]	Virtual network services	IP	Virtual private networks	Static
ANTS[11]	Composing network services	IP	Internet protocols	Dynamic integrated
Switchware[23,24]	Composing network services	IP	Internet protocols and security	Semidynamic
Xbind[12]	Service creation	ATM	Multimedia networks	Static, open signaling
X-bone[25]	Deployment of IP overlays	IP	IP overlays	Semidynamic
Smart packets[26]	Network management	IP	Managed nodes	Dynamic, discrete
NetScript[27]	Composing network services	IP	VANS	Dynamic, discrete

CANEs[28]	Composing services	IP	Composable services	Dynamic
VAN[20]	VPN creation	IP	Managed nodes	Dynamic
ANN[21]	Protocol deployment, application processing	ATM	ATM ports	Dynamic
Phoenix[29]	Creation of new services	IP	Core network nodes	Dynamic
Magician[30]	Creation of new services	IP and wireless networks	Nodes	Dynamic
Programmable framework[7–9]	Programmability of schedulers	IP	Nodes (scheduling)	Dynamic

on top of heterogeneous networking platforms. It attains interoperability across these platforms by adopting Common Object Request Broker Architecture (CORBA) that provides open access to BIBs.

4.2.1.2 The Extended Reference Model

The binding model is implemented as one of the layers in the extended reference model (XRM). The XRM consists of three layers: the broadband network, the multimedia network, and the services and applications network. Figure 4.1 shows the three layers of the XRM. The broadband network is used to physically transport the cells. It consists of ATM switches, high-speed communication links, and multimedia end devices. The layer above the broadband network is the multimedia network. This network provides middleware services to the applications by providing open interfaces to give access to the BIB and implement end-to-end QoS guarantees. This is achieved by deriving QoS abstractions from the broadband network. Service abstractions represent states of services. These abstractions are utilized by the services and application network for managing and creating new services through dynamic composition and binding.

Each of the three layers of the extended resource model is further divided into five planes. The N plane is responsible for system management and consists of monitoring and control of individual states. These states represent the status of a link, the temperature of the interface card, etc., and house managed objects in the management information base (MIB). The M plane looks after the resource functionality in terms of buffer management and link scheduling. Flow control is another example of resource control. It operates at the cell/frame level. The D plane models the main network component switches, multiplexers, as a global distributed memory. For example, communication links are modeled as FIFO memory and switches and processors as random access memory. The C plane facilitates exchange of state information between distributed buffer management and link scheduling. The U plane describes the adaptation layer protocols. This consists of the ATM layer and ATM adaptation layering functions.

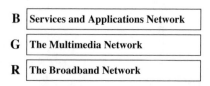

Figure 4.1   Three layers of the XRM.

The two interfaces are that between R and G and that between G and B. The R and G interface abstracts the states of resources in the broadband network. These resources are end nodes, devices, switches, links, processors, and their respective capacities and bandwidths. The open access through these interfaces allows the states of these resources to be manipulated, managed, and monitored by a remote node. These abstractions appear as a collection of interfaces called BIBs. The G and B interface is called the Broadband Network Service (BNS). It provides the following services:

- Connection management, routing, and admission control
- Virtual circuit services
- Virtual path services
- Virtual network services
- Multicast services

Like BIB, BNS uses these services to create high-level services.

4.2.1.3 The Service Creation Process

The service creation process includes the following steps:

1. A service skeleton for an application such as a virtual circuit or path is created. The skeleton structure is based on the graph from a source node to the destination node.
2. The skeleton is mapped into the appropriate name and resource space and a new network application is created.
3. Next, a media transport protocol is associated with the application and a transport application is created.
4. The transport application is associated with the resources, and thereby a network service is created.
5. Lastly, a service management system is associated with the network service and a managed service is created.

4.2.2 ANTS: Active Network Transfer System

The Active Network Transfer System (ANTS) project was developed at MIT.[11] The aim of the project was deployment of various communication protocols in active networks, thus providing options of core services, including support for the transport of mobile code, loading of code on demand, and caching methods.

These core services permit network developers to introduce new or extend existing network protocols. The ANTS creates a network programming environment by specifying the routines to be executed at the modes that forward their packets. These routines are automatically deployed at the necessary nodes through mobile code techniques. This technique reconfigures the nodes on the fly without stopping them.

4.2.2.1 Architecture of the ANTS

The ANTS network is like a conventional wide area network in which some of the routers are replaced by active nodes. Applications exchange special kinds of packets called capsules over the network. Figure 4.2 shows the architecture of the ANTS.[11]

The ANTS has three main parts:

- Capsules
- Active nodes
- Code distribution system

Capsules are like packets that contain application data and code or reference to a code to execute that data. Active nodes are the nodes on which this code is downloaded and processed. A code distribution system makes sure that codes are automatically transferred to the nodes that require them.

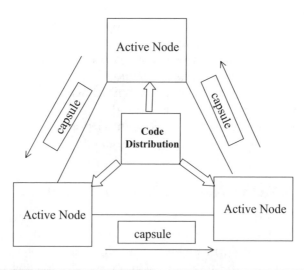

Figure 4.2   The Architecture of ANTS.

Source address	Destination Address	TTL	version	type	Previous address	Type dependent-header fields	Payload

Figure 4.3   The structure of a capsule.

4.2.2.1.1   Capsules

Figure 4.3 shows the structure of the capsule. It contains the regular IP header plus extension fields specific to the ANTS. The type field is an identifier that shows the associated protocol and forwarding routine. This is also responsible for the integrity of the code. Security is ensured by the MD5 hash function. Because the identifier is generated from the protocol contents, there is no need for a central authority to allocate type identifiers. This is in contrast to the standard protocol approaches (Open System Interconnect (OSI) and Transmission Control Protocol/Internet Protocol (TCP/IP)).

The remainder of the header contains fields common to all capsules, including version information and addresses used by the code distribution system. The fields that vary with function to be performed by the capsule are type dependent. For example, the congestion notification field in the header of the capsule indicates whether or not congestion has been encountered so far. The payload information, as in IP, contains higher-layer information that is conveyed across the network and exchanged between applications.

4.2.2.1.2   Active Nodes

In ANTS, each node provides a set of functions that are used to design forwarding algorithms. Each node executes forwarding routines within an execution model that controls the resources forwarding routines can access. The main consideration in the design of a programmable network is the safe execution of forwarding routines and efficient allocation of resources among them.

To fulfill these requirements, active nodes provide an opportunity to execute protocols within a restricted environment. This limits the protocol access to shared resources.

Node Primitives — The node primitives in ANTS as specified by Tennenhouse et al.[11] are of three types.[33]

Environment calls: Send information about the node environment, e.g., its address. Other examples of environment calls could be the information about the forwarding routines, etc.

Storage calls: Use the soft state of application-defined objects of other capsules. To speed up things, the node caches used objects. The nodes are kept updated by removing out-of-date objects.

Control operations: Direct the flow of capsules processing to other sections of the network. They are used to send capsules to other nodes by the shortest route when there is congestion in some section of the network.

The Execution Model — The execution model supports generalized packet forwarding, rather than a distributed computing environment. The model has four main properties:[33]

Fixed forwarding routine: The forwarding routine is decided at the sender; when the capsule is sent into the network, this routine does not change as the capsules travels through the network. This creates a secure environment because an intruder cannot attack another application's capsules.[33]

Selective execution: Not all the nodes are active nodes in the sense that they may not execute a particular forwarding routine; some of the nodes may not execute the forwarding routine depending on the availability of resources and security policies. These nodes are passive nodes (ordinary IP router).[33]

Resource limits: Due to the security concerns, the forwarding routines are restricted to resources they are allowed to use when they run. They make sure that forwarding routines are completed as soon as possible. They also restrict the usage of bandwidth and memory.[33]

Protocol-based protection: This characteristic of the execution model makes sure that only capsules belonging to the same protocol can share the state. The protocol determines which data can be accessed by a capsule while in the network. This mechanism ensures that capsules belonging to protocols may not intrude upon each other.[33]

When a capsule arrives at a node, its forwarding routine is executed. The routine can use the capsule's header fields and payload as it executes. It also completes some other tasks, such as sending the capsule to a remote node. The forwarding routine also enters any information into the soft state that is needed at the node for future capsule references belonging to the same protocol. During the execution of the forwarding routines, the nodes monitor their integrity and handle any errors that arise. To void a routine from corrupting an entire node, safety mechanisms of mobile code technologies are executed. Traditional operating system protection schemes, on the other hand, are too heavyweight for this because they are too slow to keep up with the forwarding rate of the capsules.

The active nodes have the capability to limit the resources that a forwarding routine can use when it executes. This prevents one protocol from consuming resources of other protocols. Each active node has a robust monitoring system, which makes sure that forwarding routines are not caught in loops, they do not send too many capsules, or they do not leave too much of the soft state. Inside the network, infinite forwarding loops are broken using the IP time-to-live (TTL) field. The functionality of the TTL field is the same as it is in IPv4. Each time a capsule passes through an active node, it decrements the value of this field by 1. If this value after being decremented becomes zero, the capsule is discarded.

A very important issue is the protection of network resources. A badly written forwarding routine would consume a large portion of the network's resources. For example, a badly written forwarding routine would saturate network links if it causes a capsule to unnecessarily travel across the network links many times. ANTS solves this problem by having a trusted authority certify forwarding routines.

4.2.2.1.3 The Code Distribution System

The code distribution system distributes forwarding routines to nodes where they are supposed to run. Even though ANTS will accept many new protocols in the future, having protocols support all types of traffic is a challenging task and difficult to achieve. Because there has been progress in the area of interoperability, there is a possibility that protocols would support all traffic types.

In ANTS, forwarding routines are transferred separately from the capsules they are associated with and are cached at nodes. At network's edges, applications provide forwarding routines to their local node before sending the corresponding capsules into the network. Within the network, a lightweight routine transfers the code along the path of the capsule if it is not already cached. The advantage of this caching is that applications can send capsules any time without first setting up a connection.

Network congestion causes the code loading to fail. In this situation, ANTS considers the capsules that triggered the load as lost. TCP in this case provides the appropriate degree of end-to-end reliability.

4.2.2.2 Programming

ANTS protocols can directly utilize a network's topology and load information. Because of this flexibility, network designers and operators can construct novel types of routing, flow setup, congestion handling, and scheduling protocols.

ANTS can be used for multicasting. Multicasting, in which one host communicates with many other hosts by replicating messages, has been the focus of research since OSI and TCP/IP were presented. In fact, the Internet has been upgraded to support multicast for applications like video. The ANTS multicasting is simple, but works for networks with a mix of active nodes, IP routers, and asymmetric routes.

Figure 4.4 shows the interaction of two types of capsules: subscribe capsule and multicast data capsule. The subscribe capsule is sent periodically by applications that want to receive messages sent to the group. It travels across the network to a particular sender. Along the way, it refreshes soft-state forwarding pointers that are kept at nodes to form a distribution tree. The multicast data capsule delivers a copy of a message to each group member. It simply routes itself along the distribution tree by using the pointers it finds in the soft store, spawning new copies of the message as needed. The subscribe capsule begins by looking up the forwarding pointers for the group

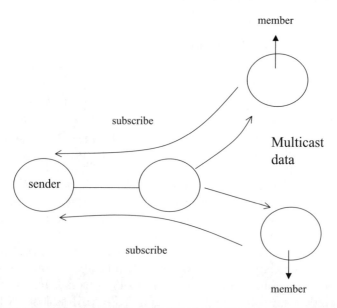

Figure 4.4 Subscribe capsule and multicast data capsule.[33] (© 1999, IEEE)

in the node's soft store, creating a new pointer if none are found. To separate the forwarding pointers of different groups in use simultaneously, the protocol stores the pointers under a key that combines the group and sender addresses. Once the capsule locates the pointers, it adds a new entry that points in the direction from which the subscribe capsule has come, if such a pointer is not already there. The capsule then continues toward the sender if it is time to refresh upstream pointers.

The multicast data capsule travels through the network by following the forwarding pointers. When a copy of the capsule reaches its destination, an empty set of forwarding pointers signifies that it is to be delivered to a local application that is a member of the group.

Together, these capsules provide a service with the central property of network-based multicast and efficient use of bandwidth. The service differs from IP multicast in two significant respects. First, because it is local to the nodes that use the protocol, multicast-capable routers need not be separately identified or organize themselves into a tree. Second, it provides a different multicast primitive, because members subscribe to the combination of a group and sender. This choice is typical of the flexibility that ANTS offers. If applications need multiple senders, they can form multiple distribution trees by having members subscribe to each sender. Alternatively, they can consider the sender as the root of a core-based tree, and route capsules up the tree toward the root and then down other branches.

The ANTS toolkit is implemented in Java, as a stand-alone network system rather than an extension of an existing IP implementation, and transfers forwarding routines in Java's classfile format. Java's flexibility as a high-level language and its support for dynamic linking/loading and multithreading facilitate the implementation of the architecture.

4.2.3 Switchware

Switchware developed at the University of Pennsylvania[23,24] uses three important components: active packets, switchlets (active extensions), and a secure active router infrastructure. Active packets are similar to MIT capsules. Active packets carry programs consisting of both the code and data and replace both the header and payload of conventional packets. Basic data transport can be implemented with code that takes the destination address part of its data, looks up the next hop in a routing table, and then forwards the entire packet to the next hop. At the destination, the code delivers the payload part of the data. Switchlets or active extensions are dynamically loadable programs that provide specific services on the network node. Active packets are programmed in

Programming Language for Active Networks (PLAN). PLAN programs are strongly typed and statically type checked to provide safety before they are injected into the network. Additionally, PLAN programs are made secure by restricting their actions, such as the manipulation of node-resident state. To avoid the limitations of PLAN programs, active packets can call switchlets, which are dynamically loaded onto routers and provide services to active packets. Switchlet modules are written in collaborative application markup language (CAML), which supports formal methodologies to provide security at compile time, and no interpretation is needed. The switchlets are subjected to heavier security checks than active packets in PLAN. They are statistically type checked on arrival at a router. To realize security of routers, the secure active network environment (SANE)[24] was designed at the University of Pennsylvania. This ensures security of the entire system. SANE identifies a minimal set of system elements (e.g., a small area of the BIOS) on which system integrity depends and builds an integrity chain with cryptographic hashes on the image of the succeeding layer in the system before passing control to that image. If the image is corrupted, it is automatically recovered from an authenticated copy over the network.

Switchware architecture provides a very secure infrastructure for active networking, but there are certain problems associated with it. First, the PLAN programs are not powerful enough for many applications; therefore, switchlets have to be installed *out of band* to provide "handles" for the PLAN programs. This makes the system less dynamic. Second, as the active packets are statically type checked, this makes the proceedings slower than dynamic type checking, which allows the runtime error checking, thus making it fast. Furthermore, if, in spite of static type checking, an error occurs, it would make the debugging more challenging, because the active packets and switchlets are executed remotely from the programmer. Third, use of two different languages, PLAN and CAML, could make the architecture slow and complicated.

4.2.4 Smart Packets

Smart packets are another capsule-based approach.[26] They aim to improve the performance of large and complex networks by leveraging active networking technology without placing an undue burden on nodes in a network. Smart packet programs are written in a tightly encoded, safe language specifically designed to support network management and avoid dangerous constructs and accesses. Smart packets are used to improve the management of large complex networks by (1) moving management decision points closer to the node, (2) targeting specific aspects of the node for information, rather than collection via polling, and (3) using programming for management, allowing fast network control.

Currently, network management is performed by having management stations routinely poll the managed devices looking for anomalies. As the number and complexity of nodes increase, management centers become points of huge "bursts" filled with large amounts of redundant and old information from nodes. A further problem with polling is that a component can suffer multiple state changes in less than one round-trip time and, indeed, can oscillate per packet. It is important that network management employ techniques with more immediate access and more ability to scale.

Management centers can then send programs to the managed nodes to gain three advantages. First, the information content returned to the management center can be tailored in real-time to the current interests of the center, thus reducing the back traffic as well as the amount of data requiring examination. Second, many of the management policies employed at the management centers can be embodied in programs that, when sent to managed nodes, automatically identify and correct problems without requiring further intervention from the management center. Third, smart packets reduce the monitoring and control loop measurement time. Control operations are taken during a single packet traversal of the network, rather than through a series of SET and GET operations from a management station.

The smart packet consists of a packet header followed by a payload. It is encapsulated within an Active Network Encapsulation Protocol (ANEP), which in turn is carried within the IPv4, IPv6, or User Datagram Protocol (UDP).

A complication in implementation of smart packets is that the IP router does not have a notion of a datagram whose contents are processed at intermediate nodes. The IP router simply examines the datagram header and forwards the datagram. For smart packets, however, the router must process the contents of the datagram before forwarding it. Another complication is that the router should check the contents of the datagram only if the router supports smart packets; otherwise, the router passes the datagram through. To deal with this problem, an IP option called *router alert* has been modified by the authors of this project. Two programming languages have been developed in the smart packets project: Sprocket, which is high-level language with built-in features to support network management, but with security-threatening constructs such as pointers removed, and Spanner, a complex instruction set computers (CISC) assembly language, into which Sprocket is compiled. The network API is implemented by a Spanner virtual machine implemented by a daemon running in an active node. The virtual machine addresses the security issues, but raises the concerns regarding time and resources. The verification process (e.g., use of public key certificates) raises issues about the size of the entire smart packet. The summation of the sizes of IP, ANEP, the smart packet header, and the certificate and subtraction from the size of the maximum-length IP datagram over the Ethernet leaves only 1024 bytes for the program execution. Another concern is that many of the techniques that reduce the time needed to process the certificate require additional certificates in the message. This further reduces the bytes for the program execution. For this reason, smart packets allow unauthenticated programs to execute, but in a limited environment.

4.2.5 Netscript

The Netscript at Columbia University is a middleware to facilitate programming of intermediate network nodes.[27] The Netscript programming language allows script processing of packet streams with a focus on routing, packet analyzers, and signaling functions. Netscript programs are organized as mobile agents that are dispatched dynamically to remote systems and executed under local or remote control. Netscript programs are message interpreters that operate on streams of messages. Messages can be encoded either as high-level Netscript objects or in a format compatible with existing standards. The appropriate agents to accomplish the desired functionality of the protocols process packet streams arriving at intermediate nodes. The Netscript agents dispatched to intermediate nodes allocate resources to the virtual network and configure its processing functions to handle routing and flow control of packet streams over the network. The Netscript project consists of three parts: an architecture for programming networks at large, an architecture of a dynamically programmable networked device, and a language called Netscript for building networked software on a programmable network. A Netscript network is viewed as a collection[27] of virtual network engines (VNEs) interconnected by virtual links (VLs). A VNE can be programmed by Netscript agents. Netscript agents can be dispatched from any VNE to any other VNE. The collection of VNEs and VLs defines a Netscript Virtual Network (NVN).[27] Netscript provides a language to program an NVN. An NVN may correspond loosely to the underlying physical network. A physical network node may be responsible for executing multiple VNEs, and VLs may correspond to a collection of physical links that interconnect VNEs. Additionally, a VL can interconnect any number of VNEs; this is particularly helpful for handling broadcast links.

A VNE consists of three layers:

1. Agent services layer, which provides multithreaded execution to support delegation, execution, and control of agent programs. It also provides message communication services among agents.
2. Connectivity services layer, which is responsible for interacting with the underlying physical environment to allocate and maintain VLs to neighboring VNEs. It provides a library of primitives used by Netscript programs to control the allocation of VL resources, scheduling, and transmission of packets over VLs.
3. Netscript interpreter, which provides a multithreaded execution environment for Netscript agents. It maintains local libraries of agents and provides access to a distributed global directory of agents executing at different VNEs.

Two models of Netscript have been proposed to interoperate with existing networks and protocols. In the first, Netscript would form a glue layer over the existing technology. For example, a VL of Netscript may be implemented on top of IP links or tunnels, or use an ATM virtual circuit. In the other model, IP packets can be encapsulated in a Netscript header and routed by the Netscript program to neighboring VNEs, where they are processed again by the IP routers. Netscript is a dynamic language. This means that programs or devices can be added or removed from a VNE on the fly at runtime without disturbing the execution of existing protocols. When a Netscript program arrives at a VNE for execution, the environment automatically configures the program and connects it to other programs based on the type signature of that program. This automatic configuration is important because the internal state can differ radically from one VNE to another. If a program depends for its execution on other programs or protocols (e.g., IP) that do not currently reside on a given engine, the program simply remains dormant until the dependent programs arrive from the network. Autoconfiguration saves the programmer from writing complex, tedious, and error-prone scaffolding code to ensure that a program will execute on all VNEs.

The key difference between Netscript and ANTS and the Switchware is that Netscript views the network as a single programmable abstraction (the NVN) rather than a heterogeneous collection of programmable routers, switches, and end nodes. Netscript does not explicitly address the efficiency of node programs to handle fast real-time requirements and the security of programmable nodes against the unauthorized access.

4.2.6 CANEs: An Execution Environment for Composable Services

Composable active network elements (CANEs) is an execution environment (EE) specifically built for composing services within the network.[28] APIs enhance the ability to *compose* services from building blocks (or *components*). Viable active network APIs must contain a composition mechanism that can be used to create a composite service from components, eliminating the need to build services from scratch.

The nodes in CANEs are connected via channels. Channels are used to transport packets between nodes. Each node repeatedly removes packets arriving on incoming channels and processes the packets. The packet processing is dependent on the information carried in the packet and the current state of the node. Processing at the node is comprised of two parts: a fixed part that represents the uniform processing applied to every packet, and a variable part that depends on the packet contents and node state. The fixed part corresponds to the virtual machine that is exported by the network architecture; the variable part is encoded using the network API and is input to the node virtual machine. Depending on the facilities provided by the network API, node behavior can be tailored by providing suitable input to each node virtual machine on the network path.

The fixed part of the node behavior is called the *underlying program* running at each node. User-specific functionality that forms the variable part of packet processing is called the *injected program*. Users inject programs into the network using the network API. Mechanisms to inject user code into the network must address the issues of the form and transport of the injected code to specific nodes in the active network. Note that users of the network are not allowed to install new underlying programs. Therefore, the form of the injected program and the interaction between the underlying and injected programs essentially define the programming interface supported by the network. To reason about the composite service offered by the network, the programming interface must formally express the interaction between the underlying program and the injected program; i.e., the programming interface must define how the control flows between the injected and underlying programs, and how the state is shared between the two programs.

In CANEs, the programming interface consists of identifying choices the underlying program should make at well-defined points in its computation. In effect, the underlying program offers a *menu* from which the user can select customized processing, which should occur inside the network.

CANEs supports customization and programmability of network services by putting together underlying and injected programs, and is able to

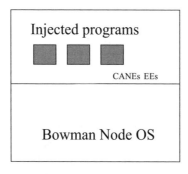

Figure 4.5 CANEs implementation.

derive properties of this combination from the properties of the components, that is, the underlying and injected programs. A satisfactory solution should not only support dynamic injection of programs, but also allow more complex services to be built up from simpler ones, and encourage modularity and abstraction.

Figure 4.5 shows a thread-level abstraction of CANEs implementation, which runs above the Bowman[34] node operating system. The figure shows a high-level view of processing at an active node.

Upon start-up, the CANEs EE spawns a control Bowman process (called an a-flow) that contains a thread for handling signaling messages. Underneath the EE, system threads provide support for packet input, output, timers, and other NodeOS routines. Each user's flow is executed in its own process and can spawn multiple independent threads that share an address space. The CANEs EE itself provides a library that implements the CANEs API and resides in the system as a single EE process that contains a set of threads to handle signaling and timers.[28]

Both the underlying and injected programs are dynamically loaded using the underlying OS's extension mechanism.

4.2.7 Supranets

Supranets[22] are virtual networks that are built on top of the Internet. This is achieved by inserting a supranet layer between the traditional transport and network layers in the Internet architecture. Supranet has its own address space. Supranet addresses are assigned to supranet hosts and routers by the creator. The virtual network components are mapped onto physical components. Thus, a physical host that is also a supranet host also has two addresses, a physical one and a virtual one. The routing in supranet is fixed; i.e., to reach supranet host C from host A, packets would always visit B first. For interoperability, supranet packets are encapsulated into IP packets. Tunneling is therefore an important technique in the construction of supranets. Because supranet membership is restricted, it is possible to build multicast trees that define the paths from each member of the group to others. When building such paths, appropriate mechanisms exist that can be used to minimize the overall group traffic. Every change in group membership requires a modification of all trees pertaining to the multicast group. Additional security mechanisms, such as private keys for each group, are used to provide security services at the multicast group level.

The security mechanism in supranet is based on asymmetric key cryptography. To authenticate its messages, the sender encrypts a portion of the supranet header with its private key. The receiver uses the public key of the sender for the decryption. Because the sender is the only owner of its private key, that particular sender must have necessarily generated any messages encrypted with this key. The checksum field part of the supranet header is encrypted, to guarantee that no outsider has changed the contents of the packet. An advantage of this scheme is that it allows the receiver to automatically detect transmission of corrupted packets. However, a drawback is that this is normally done at a higher layer, also by the TCP, thus resulting in a duplication of functions and efforts. If a user asks for an unreliable service, e.g., by selecting the UDP at the transport layer, the supranet layer would still need to compute the value of the checksum field. This would cause a limitation in the case of audio and video streams that can tolerate a certain amount of errors, but can hardly tolerate any transmission delays.

4.2.8 Switchlet-Based Tempest

Tempest[13–16] is a programmable switch network with an open signaling functionality. It creates multiple coexisting control architectures in broadband ATM environments. In Tempest, a given ATM switch is simultaneously controlled with multiple controllers by partitioning the resources of that switch between those controllers. Such partitions are named *switchlets*. The switchlet is a complete switch. The set of switchlets that a controller or group of controllers possess forms its virtual network with exclusive access to its subset of the physical network resources. The virtual private networks (VPNs) are dynamically created and assigned a dedicated set of network resources (virtual path identifier (VPI)/virtual circuit identifier (VCI) space, buffer space, a subset of switch ports, and bandwidth). When the VPN is devolved, the assigned resources are released and given to other control architectures. The switchlets allow new control architectures to be introduced dynamically into an operational network. The framework enables the end users to dynamically associate code with network

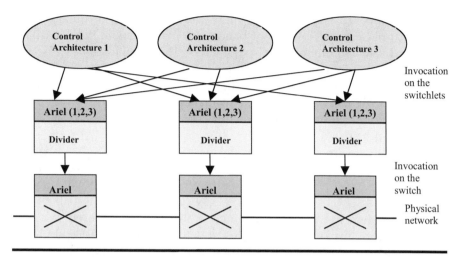

Figure 4.6 Multiple control architectures operating simultaneously.[14] (© 2001, IEEE)

resources in running the control architecture, to achieve application-specific control. The multiple control architectures operate simultaneously on a single physical network. Resources in an ATM network are divided by using two software components: a switch control interface called Ariel and a resource divider known as Prospero. Prospero communicates with an Ariel server on an ATM switch, partitions the resources, and exports a separate control for each switchlet created.

Figure 4.6 shows distinct control architectures sharing the resources of the physical network. Examples of control architectures are Ipsilon IP switching, ATM forums UNI/PNNI, Multiprotocol Label Switching (MPLS), etc. Each control architecture makes invocations on the Ariel interface exported by the divider for the corresponding resource allocation on that switch. The tying of switchlets is performed by an entity called network builder. The IP over ATM network has been used as the bootstrap virtual network, and the distributed proactive environment has been provided by CORBA. Resource management in Tempest is done at two levels: at the individual call or flow level within virtual private networks (VPNs) and at the level of the VPNs. The first issue is whether a new VPN can be created given the current state of the network, a situation referred to as virtual network admission control. Bandwidth allocated to a VPN can be guaranteed for the duration of its existence. Alternatively, a virtual network can be given statistical or soft guarantee, and the available multiplexing gain resulting from the statistical multiplexing of switchlets can be exploited to allow new VPNs to be created. The admission control in Tempest is based on estimation of the effective bandwidth of the arrival processes at a queuing system, which is computed from online traffic measurements. This approach of admission control is very flexible and equally applicable to all levels of bandwidth management, but it requires users to know large numbers of parameters. These parameters are sometimes difficult to obtain, so users are forced to allocate resources based on peak rate. An alternative might be to use the traffic parameters such as peak and mean rates and delay variation tolerance; however, these give insufficient information to achieve a useful degree of statistical multiplexing gain.

Security issues regarding the code injection are not clearly discussed for this architecture. The client is permitted to add code to the management server (e.g., Simple Network Management Protocol (SNMP) agent) running on the switch. The client code becomes an intrinsic part of the switch management interface for the client. This allows different users to customize their management policies and functions, but the code must be very safe and type checked before it is used. Because the code is injected out of band, Tempest provides no automated on-the-fly upgrading functionality. Hence, it cannot be categorized as an active packet network, because it deals only with the dynamic control of networks.

Tempest tries to find an optimal VPN topology, which is a time-consuming process. In a highly dynamic VPN service, the performance of the process used for automated VPN creation is crucial. The time taken to determine the topology and resource allocation of a new VPN can even inhibit the deployment of lightweight VPN for a brief period. The switchlet approach tries to find the best VPN topology, which is computationally expensive because of the large number of possible VPN topology candidates. The problem becomes more complicated if certain QoS requirements are also required by the customer, e.g., maximum delay variation along the paths to the receiver.

Tempest-based VPNs use the *resource revocation* algorithm for the reallocation of resources among the connections. The resource revocation algorithm reallocates the bandwidth among the connections in case of congestion in the network. It takes bandwidth from some connections and gives it to other connections. This approach can leave some of the connections starved of bandwidth and with higher end-to-end delays.

4.2.9 Routelet-Based Spawning Networks

Routelets apply the concept of resource partitioning to the internetworking layer. The spawning networks or routelets project at Columbia University is based on the Genesis Kernel.[17,18] The Genesis Kernel has the capability to spawn child network architectures that can support alternative distributed network algorithms and services. The function of the Genesis Kernel is that of a resource allocator, arbitrating between conflicting requests made by spawned virtual networks. Virtual networks created through spawning by parent networks inherit architectural components from their parent networks. Thus, a child network can be parent to its own child networks, creating the notion of nested virtual network architectures.

At the lowest level of the framework, a transport environment delivers packets from source to destination through a set of open programmable virtual router nodes called *routelets*. A routelet comprises a forwarding engine (IPv6, MPLS, cellular IP), a control unit (spawning, composition, allocation, and data path controllers), and a set of input and output ports, as shown in Figure 4.6. A virtual network is characterized by a set of routelets interconnected by a set of virtual links, where a set of routelets and virtual links collectively form a virtual network. Routelets process packets along a programmable data path at the internetworking layer, while control algorithms (e.g., routing and resource reservation) are considered to be programmable using the virtual network kernel. Each virtual network kernel can create a distinct programming environment that supports routelet programming and enables the interaction between distributed objects that characterize the spawned network architecture. The programming environment comprises a metabus (CORBA based) that partitions the distributed object space supporting communications between objects associated with the same spawned virtual network. The metabus (middleware support) is a per virtual network software bus for object interaction. The programming environment also consists of a binding interface base, which supports a set of open programmable interfaces on top of the metabus, providing open access to a set of routelets and virtual links that constitute a virtual network. Genesis Kernel supports a virtual network life cycle for the creation and deployment of virtual networks through profiling, spawning, and management. Profiling captures the specifications of the virtual network in terms of addressing, routing, signaling, security (through validity checks), control, and management requirements in an *executable profiling script*, which is used to automate the deployment of programmable virtual networks. Based on the profiling script, the process of spawning sets up the topology and address space, allocates resources, and binds transport, routing, and network management objects to the physical network architecture, thereby dynamically creating new virtual networks.

The nesting property of routelets allows a child network to inherit life-cycle support from its parent. A child routelet is bootstrapped by a parent spawning controller. The spawning controller interacts with the allocation controller to reserve a portion of the parent routelet's computational resources for the execution of the child routelet. Next, the child routelet state is initialized. During this phase, a spawner acquires all the required transport modules unavailable in the parent network. When the initialization

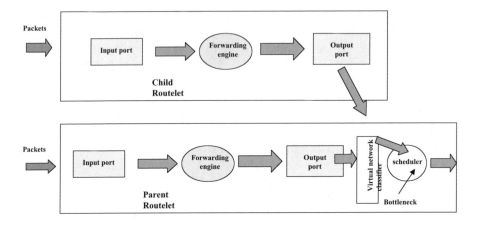

Figure 4.7 Packet scheduling in nested routelets.[17] (© 1999, IEEE)

of the routelet's state is complete, the child control unit is spawned. During this phase, the standard controllers are created, e.g., spawning, composition, and allocation controllers. Finally, the child network's data path is composed and its queues are configured to forward traffic to parent network queues. The last phase of spawning the child transport system is the binding of virtual links to a set of distributed routelets forming a virtual network topology.

Spawning networks require more computational resources for routing and congestion control due to their complex programming architecture, thus reducing transmission throughput. Profiling and spawning of VPNs becomes very slow when new routelets are created to satisfy communication needs of distinct clients. The performance of packet forwarding could also be affected due to the hierarchical link-sharing design for routelet nesting and virtual network isolation. This project does not address the bandwidth allocation issues among different services. Figure 4.7 shows the path of packets sent to the scheduler. All the child routelets send packets to the scheduler of the parent routelet, creating a bottleneck at the output of the parent routelet, and thus incur higher end-to-end delays for services. This work also addresses the important issue of searching for the best path for the service.

4.2.10 Hierarchical Fair Service Curve Scheduling in Darwin

The Darwin[19] project suggests customizable resource management policies for network management. The resources that are managed dynamically are links, switches, capacity, storage capacity, scheduling, and computation resources. Architecturally, the Darwin framework includes a service broker

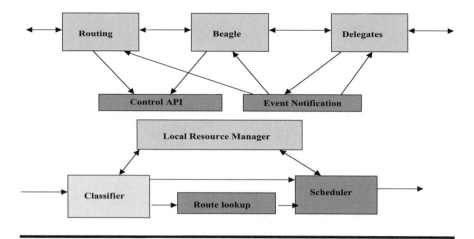

Figure 4.8   Darwin-based router architecture. (© 2000, IEEE)

called Xena that maps user requirements to a set of local resources. The resource managers communicate with Xena using the Beagle signaling protocol. Darwin uses a hierarchical fair service curve (H-FSC) as the scheduling mechanism, as shown in Figure 4.8.

The aim of H-FSC is to provide guaranteed QoS for each class by using nonlinear service curves, to allow priority (decoupled delay and bandwidth allocation) among classes and to properly distribute excess bandwidth. The scheduling in H-FSC is based on two criteria: a real-time criterion that ensures that the service curves of all leaf classes are guaranteed, and a link-sharing criterion that aims to satisfy the service curves of interior classes and fairly distribute the excess bandwidth. The real-time criterion selects the packet only if there is a chance that the service guarantees for leaf classes will be violated. Otherwise, the link-sharing criterion is used. In H-FSC, each leaf class maintains eligibility time, deadline time, and virtual time associated with the packet at the head of a class queue, and the interior classes maintain only virtual time. The deadlines are used to guarantee service curves of leaf classes. The eligibility times are used to arbitrate which one of the two scheduling criteria (real-time or link sharing) to use for selecting the next packet. At any given time when there is no eligible packet, the algorithm applies the link-sharing criteria recursively, starting from the root and stopping at a leaf class, selecting at each level the class with the smallest virtual time. In H-FSC, virtual times are used by the link-sharing criterion to distribute service among the hierarchy according to classes' service curves. The link-sharing criterion is used to select the next packet only when the real-time criterion is not used. If $E(t)$ is the minimum amount of service that all sessions should receive by time t, such that the aggregate service time required by all sessions during any future time interval (t_1,t_2) cannot exceed $C^*(t_2 - t_1)$, where C is the server capacity,[19] then

$$E(t) = (\text{min serv of act sess})_t + [\max[(\text{serv of act sess}) + (\text{serv of pass sess})] - C^*(t_2 - t_1)] \tag{4.1}$$

where the first part of Equation 4.1 shows the minimum amount of service received by active sessions by time t_1. The first term in the second part represents the maximum amount of service required by the sessions over the interval (t_1, t_2) that are already active at time t_1, and the second term shows the maximum amount of service required by the sessions that are passive up to time t_1, over the same interval. Equation 4.1 represents the worst case, when all active sessions continue to remain active up to time t_2 and all passive sessions become immediately active at time t_1 and remain active during the entire interval (t_1, t_2).

The major problem with the implementation of this algorithm is efficient computation of $E(t)$. $E(t)$ depends not only on the deadline curves of the active sessions, but also on the deadline curves of the passive ones. Because the deadline curve of a session depends on the time when the session becomes active, the algorithm needs to keep track of all possible changes, which in the worst case is proportional to the number of sessions. Second, even if all deadline curves are two-piece linear, the resulting curve $E(t)$ can be n-piece linear, which is difficult to maintain and implement efficiently.

H-FSC uses approximations to estimate $E(t)$. For example, in a typical situation when a session i becomes active, and the sum of slopes of deadline curves of all sessions served by the scheduler at that particular time is greater than the server's rate, it becomes impossible to satisfy the service curves of all sessions. The solution suggested by the authors is the allocation of an overestimated high value of $E(t)$ to satisfy the service requirements of all sessions; i.e., the server allocates enough service in advance to all the sessions such that it has sufficient capacity to satisfy the bandwidth requirements of all the sessions when a new session becomes active. This can lead to a violation of QoS guarantees of real-time services and also unfairness among the services.

4.2.11 Virtual Active Network (VAN)

In the virtual active network (VAN) framework,[20] VPNs are provisioned on demand within an active network infrastructure. This allows customers to deploy and manage their own active services by using active packets to install or upgrade service-specific functionality, without interaction with the VAN provider. Each VAN has a dedicated set of resources, including link bandwidth, node processing, and memory resources, which is obtained through a programming interface. This work of VAN is conceptually very similar to that of Tempest, except it uses active packets for service creation and provisioning. In this framework, the VAN provider's management interface is explicitly separated from the *VAN management interface*, through which customers can exert fine-grained control and deploy customized network services. The management interface can be realized using management protocols such as SNMP or CMIP or by using active packets. However, the project does not clearly mention how the resource partitioning is enforced. The issues of resource management and distribution are also not addressed in the project. The service management system of a VAN can install specific network services on the VAN by sending an active packet to each execution environment within the VAN. This packet upgrades the packet classifier, sets the buffers for each class, and substitutes the FIFO scheduler with a scheduler for multiclass traffic. This work does not explain how the multiclass scheduler allocates priority to the sessions and how QoS is maintained during congestion in the network.

4.2.12 Active Network Node (ANN)

The ANN project[21] team at Washington University in St. Louis devised a hardware implementation for the scalable Active Network Node architecture for gigabit environments. Their platform consists of a general-purpose central processing unit (CPU), one or two FPGAs, 64 MB of memory, and an ATM Port Interconnect Controller (APIC) chip connected on a peripheral component interconnect (PCI) bus. In addition, their Router Plug-ins[21] research software platform is used as a framework for the development of a NodeOS.

ANN[21] proposes a high-performance ATM-based Active Network Node system that provides a rapid protocol deployment and application-specific data-forwarding capability by using field-programmable gate arrays (FPGAs) and the distributed code caching technique. The Active Network Node (ANN) consists of a set of active network processing elements (ANPEs) connected to an ATM switch fabric. The switch fabric supports eight ports with a data rate as high as 2.4 Gbps on each port. Each ANPE comprises a general-purpose processor, a large FPGA, a cache, and a dynamic RAM (DRAM). The ANPEs are connected to the ATM fabric via an ATM Port Interconnect Controller (APIC) chip. The APIC is an ATM host network device with two ATM ports and a built-in peripheral component interconnect (PCI) bus interface. The APIC implements VC switching in hardware and is capable of forwarding cells directly without passing

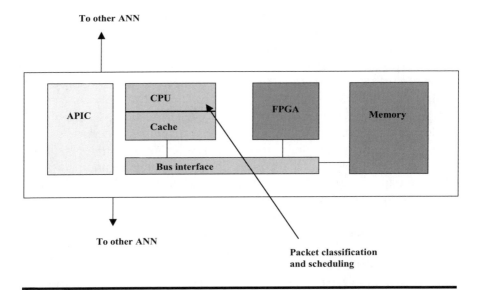

Figure 4.9 Active Network Node (ANN) architecture. (© 1999, IEEE)

them to the processing engine (processor and FPGA). The processor can take care of the majority of functions applied to a packet, while the FPGA can be programmed by the processor to implement the most performance critical algorithms in hardware. APIC can distribute individual flows to the CPU or the FPGA on a per VC basis. A packet can first go to the FPGA and then be either passed to the processor or forwarded straight through APIC to the link, depending on whether there is additional processing required. The more ANPEs there are, the less bandwidth of switch backbone that will be consumed. Figure 4.9 shows the active node architecture.

The software infrastructure consists of NodeOS and execution environments (EEs). The NodeOS implements services such as packet scheduling, resource management, and packet classification. The NodeOS offers these services to the EEs running on top of it. The EEs considered here are ANTS[11] of MIT and Distributed Code Caching for Active Networks (DAN).[35] The code blocks implementation of application-specific network functions, called *active plug-ins* in the context of the DAN[34] architecture. Active plug-ins are downloaded and installed on the node. The downloading is triggered by the occurrence of a reference in a datagram by a special configuration packet, or by an administrator. Active plug-ins create instances for packet forwarding. The packet classification and scheduling is performed by a CPU. The packet scheduler used in ANN is based on a deficit round-robin (DRR) scheduler.[36] DRR is a modified version of weighted round-robin, which handles the variable packet sizes. The main attraction of DRR is its ease of implementation. However, like weighted round-robin, it is unfair at timescales smaller than a round time of the scheduler.

4.2.13 The Phoenix Framework

The Phoenix framework[29] facilitates deployment of new network services by defining an extensible mobile agent system and a set of open, safe Java-based interfaces. It also implements a highly flexible execution environment. The mobile agents allow dynamic loading of code on a set of nodes deploying this execution environment. This code is only allowed to access necessary resources, and all access is properly authorized and authenticated.

4.2.13.1 Architecture

The Phoenix framework consists of the following main components:

- Proactive console
- Proactive environment
- Proactive services
- A mobile agent system

4.2.13.1.1   Proactive Console

Active devices are managed from a proactive console. The proactive console manages the agents and services that make up the Phoenix framework.[29] The proactive console maintains a topology database of active and legacy devices in the network. Every network device (active or legacy) is represented by a *proactive device object* (PDO).[29] A PDO contains the configuration data for proactive services supported by the active device. Network administrators utilize these PDOs to install a new proactive service on active devices and to create new mobile agents. This console is the central storage for all Java class files representing proactive services and mobile agents. Furthermore, it also consists of the security and management policy databases for the framework.

4.2.13.1.2   Proactive Environment

A proactive environment offers basic primitives for programming and managing the device. This environment is built on a Java Virtual Machine (JVM). JVM security features provide basic security support in the proactive environment. Proactive services can be installed in the proactive environment at runtime to dynamically extend the functionality of the Phoenix framework.[29]

4.2.13.1.3 Proactive Services

Proactive services are Java objects with well-defined interfaces that provide new functionality.[29] Proactive services are the methods through which device functionality is accessed or affected by mobile agents. As such, the set of services installed on a device determines exactly what functionality an agent may exercise in a device. Proactive services allow third parties to add a new functionality to an active device.

A proactive service can interact with a device locally via a native library (providing access to device resources that cannot be accessed directly from Java) or remotely through remote procedure call (RPC) mechanisms such as Java Remote Method Invocation (RMI), Microsoft Distributed Component Object Model (DCOM), and Common Object Request Broker Architecture (CORBA). Native libraries are used to provide agent access to advanced capabilities, such as packet-filtering capabilities, packet redirection, etc. Legacy devices that cannot support a proactive environment can be managed via the Simple Network Management Protocol (SNMP).

The prime example of a proactive service is congestion avoidance through analysis.[29] The process of congestion analysis is initiated by the network administrator. The administrator launches a single agent to the active device topologically closest to the congested link.[29] The agent captures a representative sampling of the traffic on the link. This sample is parsed into a bandwidth-ordered list of the sources and destinations of the traffic on the link.

Each agent then traverses the link associated with the port where the majority of its assigned source traffic is arriving. Agents continue this process until they reach a point where they cannot complete their traversal, detect a loop, or arrive at the source of the traffic. At this point, these agents report the path taken back to the proactive console. Once an administrator has reviewed the information gathered, it can take action to alleviate congestion, by either reconfiguring the physical topology of a network, applying some sort of traffic-shaping behavior to parts of the network, or provisioning new network resources as required.

4.2.13.1.4   A Mobile Agent System

Mobile agents in the Phoenix framework facilitate transport of code to nodes. Each mobile agent is a collection of objects.

A mobile agent contains a route object that determines the active devices the agent will visit during its lifetime. Each agent also carries a WorkObject (WO), which contains the programming logic specifying how to interact with a proactive environment. Associated with the WO is the policy object, which defines what the agent is to do when events are encountered. For example, the policy might specify the action to be taken when the agent visits a device but cannot execute because it cannot obtain the necessary resources (i.e., services) for execution. Each WO contains one or more WorkSheet (WS) objects that specify what the agent does at each device it visits and also includes any policies that indicate how the WS should be executed. For example, a policy may specify the conditions that deter mine whether the WS should be executed at each of the devices on the agent's itinerary.

4.2.13.2 Execution Process at an Active Device

Figure 4.10 shows the process when a mobile agent arrives at an active device. It shows how agents, PDOs, and proactive services execute under the control of the proactive environment. When a mobile agent

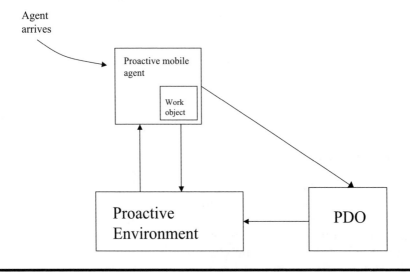

Figure 4.10   Components interaction when an agent arrives at an active device.

arrives at the active device, the proactive environment verifies the security access level and execution priority level for the agent to decide if the agent should be allowed to execute.[29] If the agent has proper authorizations and necessary resources on the device are readily available, control is passed to the agent. The agent first makes a request to the proactive environment for the interface of the PDO for the device on which it wants to operate. If the requested PDO is not available, the agent terminates or may move to another device. This is taken care of by the policy object associated with the agent WO.

If the PDO for the device is accessible, the agent then requests the PDO for the proactive services to which it needs access. The PDO creates and configures an instance of the requested proactive service and returns a reference to the agent. The agent invokes the proactive service to perform the required actions.

4.2.14 Composing Protocol Frameworks for Active Wireless Networks

A. Kulkarni and G. Minden[30] have presented a model for a protocol framework that is suitable for rapid development of any application-specific service and its deployment in a network. This active networking prototype is called Magician. This is a class hierarchy-based model that enables users to compose their own custom, flexible frameworks from either predefined or custom protocol components tailored to an application's needs.

Applications customize network resources for dynamic adaptation by injecting smart packets into the active network to modify the behavior of the active nodes.[30]

Smart packets can potentially carry code in the form of application-specific protocol frameworks composed from custom protocols.

4.2.14.1 Protocol Composition in Magician

Magician[30] is a toolkit that provides a framework for creating smart packets as well as an environment for executing the smart packets. Magician is implemented in Java. Java was chosen because it supports mobile code, serialization, and object-oriented properties such as inheritance and encapsulation. In Magician, the executing entity is a Java object whose state has to be preserved as it traverses the active network.

Magician provides a model in which an active node is represented as a class hierarchy. Figure 4.11 shows the clan-hierarchy models of an active

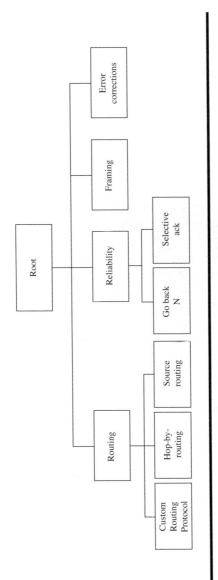

Figure 4.11 The clan-hierarchy model of an active node Magician.

node in Magician. Every protocol is derived from an abstract base protocol. Every active node provides some basic functionality in the form of certain default protocols (e.g., routing). Users may prefer to utilize these default protocols if they do not wish to implement their own schemes. To foster privacy and safety, a unique state is created for each invocation of the protocol. Providing each user with a protected copy of the state enables the user to customize his or her own state if necessary.

Users are permitted to install their own protocols. If necessary, a user creates a new protocol from scratch by extending the base abstract protocol. When the protocol is carried over to the active node by a smart packet, a new branch is created for that protocol. Alternatively, the user extends an existing protocol class at the active node and customizes it to implement application-specific behavior. In this model, the active node acts as an object whose behavior can be modified dynamically.

Magician provides two basic predefined frameworks to users:[30]

- A framework providing unreliable service
- A framework providing reliable service

The functionality of these frameworks is similar to those of the UDP datagram and TCP socket abstractions in current networks.

The basic reliable service framework adds a sequencing module, a hop-by-hop stop-and-wait acknowledgment protocol component, and a retransmission component to provide ensured delivery. A hop-by-hop protocol is chosen because in an active networking environment, the destination of a smart packet can change during its execution at an active node. This is the simplest reliable delivery mechanism that can handle dynamic destination changes. For example, if a client request packet traversing the network toward a server node encounters information about the location of a proxy that merges duplicate requests to the server, the request packet resets its destination to the location of the proxy. If the packet contains a request that has been made previously, it informs the proxy of the client's request and destroys itself. If an end-to-end acknowledgment/retransmission protocol is used in this case, it is bound to fail because the server node never receives the packet, and therefore waits for the sender to send the packet because it is next in sequence.

In Magician, users are not constrained to using only the basic frameworks. Customization of the framework is possible in many ways. Users can replace a particular functionality by replacing the protocol module implementing it. Consider the example of an active application requiring a custom routing algorithm. Active nodes in Magician implement the Routing Information Protocol (RIP) as the default, but the routing manager at the active node provides a registration feature to install custom routing protocols. A custom routing protocol is installed for an application type by registering it with the routing manager at the active node. Once registered, all smart packets belonging to the application (and sharing the same type) are routed using the custom routing algorithm. This demonstrates customization by replacement.

Users can also extend functionality of a protocol module. For example, all active nodes are addressed using the naming scheme implemented by the administrator. However, a user can choose to implement his own addressing scheme. A custom addressing protocol can extend the default addressing interface and implement a table that maps configured active node addresses into user-supplied addresses. This enables transparent translation of addresses while adhering to the basic node API.[30]

4.2.15 A Programmable Framework for QoS in Open Adaptive Environments

Active distributive management, spawning networks, Phoenix framework, and Darwin have certain drawbacks, as mentioned in the previous section. For example, the Phoenix framework provides an open interface for third parties to program network devices by using mobile agents; it also does not discuss the reconfiguration actions required when there is congestion in the network.

The future Internet is envisioned as a multiservice platform, supporting real-time traffic and complex multiuser services. The Internet model is currently in the process of being extended to include quality of service (QoS) guarantees. Due to the inherent time-sensitive characteristics of many multimedia applications (e.g., audio- and videoconferencing, multimedia information retrieval, etc.), the network is required to provide a wide range of QoS guarantees (with respect to bandwidth, packet delay, delay jitter, and loss). The guaranteed bandwidth must be sufficient to accommodate video and audio streams at acceptable resolutions, while the end-to-end delay must be small enough for interactive communication. To avoid breaks in audio and video streams' playback, delay jitter and loss must be sufficiently small. Overallocation of bandwidth reduces link (and network) utilization, and this may be a critical factor in architectures that support third-party virtual private networks (VPNs). Hence, the diversity of traffic characteristics and performance requirements of existing applications, as well as uncertainty about future applications, requires the scheduling discipline in the network elements (i.e., routers) to allocate fair delay, bandwidth, and loss rate guarantee.[7–9]

The two favored architectures for providing QoS for time-sensitive services over IP networks are Integrated Services (IntServ) and Differentiated Services (DiffServ).[8,9] IntServ requires applications to signal their service requirements to the network via reservation requests. These take the form of the Resource Reservation Protocol (RSVP). DiffServ works in the core network and employs a scalable aggregation to provide prioritization among classes of traffic. While RSVP has problems with scaling to large networks (where a large number of nodes need to maintain soft-state information on all traffic flows), DiffServ provides only a granular service that cannot offer fully optimized network (or application) performance. It is now considered that future networks will consist of combinations of these mechanisms. RSVP may be more efficiently employed in the access network, whereas DiffServ will be deployed in the higher-density backbone.[31]

The programmable framework[7-9] uses programmable networking paradigms coupled with the use of active packets, and intelligent agent technologies for the dynamic adaptation of networks. The framework also proposes different levels for the implementation of these technologies within the network.

Future networks will be heterogeneous environments consisting of more than one network operator. In this environment, agents will act on behalf of users or third-party operators to obtain the best service for their clients. In the framework, agents may be dispatched to nodes within a network and will be responsible for the maintenance of services through the virtual private networks (VPNs) created. Maintenance of a VPN may involve dynamic rerouting of connections and renegotiation of QoS targets in nodes. The renegotiation of QoS can be achieved through the reconfiguration of weights in routers spanning the VPN. The framework is based on the emerging standard for open interfaces for programmable network equipment, namely, the IEEE P1520 reference model. To date, most of the research in this area has tended to concentrate on the network generic services layer (between the U and L interfaces). An important feature of this framework is the mapping of higher-layer requirements onto actual physical resources in the network equipment.

Because of the diversity in applications, the complexity of implementation, and security issues, it is likely that the centralized network management approach will be replaced by a new distributed management system. This management architecture is called the Active Management Architecture (AMA). It can avoid scalability problems and offers flexibility to users, third-party operators, and network operators. The AMA uses all the enabling technologies mentioned. This management architecture operates at the network and element levels. The adaptive framework has the following components,[7-11] as shown in Figure 4.12:

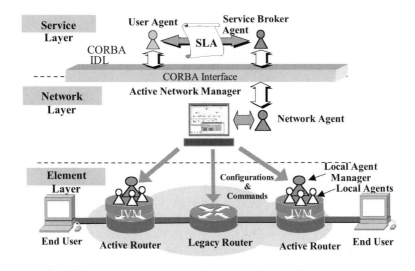

Figure 4.12 Architecture of the adaptive network.[8] (Copyright © 2004, John Wiley & Sons Ltd. Reproduced with permission.)

Agents: These are used mainly as autonomous negotiators. The *user agents* obtain a connection from a service provider. The service level agreement (SLA) contains the negotiated parameters, such as the best price, delays (which can be related to a particular scheduling scheme), guarantee of network availability, and backup routes for the connections. The *service broker agent* acts as the negotiator between the clients and the service provider. A *network agent* from the service provider negotiates with different network operators to set up a VPN.

Active network manager: This is part of the network operations and contains policy management, accounting, security, and auditing and a repository of various services. The policy server within an active manager is responsible for collecting all of this information relevant to each VPN. It then makes decisions based on the SLAs and communicates these decisions to the element layer. The goal of the policy server is to develop a response consistent with the SLA. The response is then transmitted to the network element, e.g., router, using active packets. The active network manager sends active packets for the reconfiguration of those routers within its domain. Active packets are chosen instead of conventional signaling methods such as SS7. This is primarily because conventional signaling schemes do not have any method of transporting individual local agents to the active routers, where they operate on behalf of each user agent.

Active packets: These contain codes that can be compiled at the active router, e.g., Java codes that can be compiled on the fly and executed in the JVM on the router's operating system (OS). Active packets can trigger the activation of a user agent; they can also be used to transport agents through a network.

Active router: The active router performs dynamic reconfiguration of queue weights. The network, in addition to the active routers, will contain legacy routers at the element level. The legacy routers will still depend on the traditional client/server way of performing configurations and sending commands. The active router's operating system must also support the local agents that have been dispatched from the active network manager, as well as the manager for them (local agent manager). The active router reconfigures the weights of sessions on the fly. The router serves as a means to implement the options of different classes of service according to the price and delay requirements of the clients. Agents on the routers are used as the mechanism to change the weights of services. In this architecture, JVM is proposed as the environment for agent execution, as shown in Figure 4.13. Based on the SLA, the active network manager (within each network operator) assigns weight limits to the users' VPN. This is then deployed to each active router in the VPN and operated by local agents. The agents are dispatched to the active router from the active network manager. They are sent via the active network manager because of the

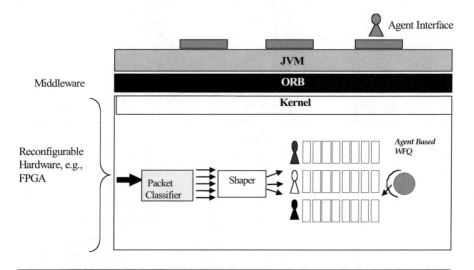

Figure 4.13   Active router architecture.[8] **(Copyright © 2004, John Wiley & Sons Ltd. Reproduced with permission.)**

security issues involved. The agents are then executed on the JVM within the router. In this scheme, JVM is proposed as the environment for agent execution. The SLA for element-level QoS contains the options of price and delays of the queuing strategy. To have a standard interface between the agents deployed on the router and the embedded hardware and software, an Object Request Broker (ORB) has been used between the JVM and the kernel. The reason for using an ORB is to provide a standard interface, because there is a wide variety of hardware and software within routers from different vendors.[7,8]

Reconfiguration of weights on the fly can be implemented in hardware, using runtime reconfigurable FPGAs. These FPGAs allow fast reconfiguration of weights on the fly without stopping the scheduler, by downloading the code from the agent to specific sections of the FPGA while the remaining sections are still functioning.

Local agent manager: This performs local routine control/management functions. It coordinates data transfer to and from different agents from different sources and clears the information data (duration, packet loss, etc.) of a session when the session ends. While a session is active, the local manager monitors resource usage and feeds back the VPN status information to the active network manager. It performs local data filtering for security purposes. The local agent manager also performs agent authentication, i.e., it identifies an agent's source before performing any task related to this agent.

Local agents: These are agents acting on behalf of each user, or a third-party service provider. Agents should be as lightweight (compact size) as possible to enable their real-time installation in the routers. These agents may be written in Java and will communicate with each other on the same node, or across different nodes, using the communication methods. These agents are used for the reconfiguration of weights of a scheduler or to adjust the weights of a connection flow (session) across a number of schedulers. An agent monitoring the queuing delay of a session in a downstream node communicates with the agent monitoring the queue delay of the same session in the upstream node to change its weight whenever the queuing delay of the session in the downstream node increases. This keeps the end-to-end delay of sessions within specified bounds.

It may be observed that the network, in addition to the active routers, will also contain legacy routers, and these will still depend on the traditional client/server way of provisioning VPNs. It is highly likely that legacy routers will not be able to support QoS targets, and in this case, it is the responsibility of the active network manager, in cooperation with the active routers, to set up specific paths for QoS-sensitive traffic flows through the network. It may also be possible for two communicating active routers spanning a legacy router to *compensate* QoS traffic flows through the legacy router, for example, by dynamic buffering and prioritization of certain traffic classes. Table 4.2 shows a typical QoS mapping scenario at three levels of the architecture. The three levels of QoS attributes correspond to the three levels of the architecture in Figure 4.12.

Deciding when to change the weights depends on a number of factors:

Traffic profile: The traffic profiles of different classes of traffic (e.g., video, voice, and data) will vary during hours of the day/day of a week, etc. Reconfiguration of weights can take place according to the traffic profile.

Price issues: Price can vary according to a network operator's tariffing policy and according to the competition between different operators. The tariffs are also differentiated (e.g., the price of different services is different, due to bandwidth and QoS targets). The weights can be modified according to the price. High-price users can have high weights, for example, for videoconferencing.

Network congestion: Network congestion will vary during the time of day and as a result of tariff changes. During congestion hours, the weights for real-time traffic can be changed.

4.3 Summary

The end nodes in a network, e.g., PCs and workstations, are generally organized as open systems whose functions can be programmed (with ease). In contrast, intermediate nodes, such as routers and switches, are closed systems whose functions are determined through long and intractable standardization processes. The review of programmable networks shows that the use of open programmable interfaces provides a foundation for service programming and the introduction of new network architectures.

Different projects that employ programmable network concepts have been discussed in this chapter. Programmable networks are classified into two communities: (1) the open signaling (OPENSIG) community and (2) the active networks community. Tempest (switchlets) and Xbind are based on the OPENSIG approach, while projects such as ANTS, smart packets,

Table 4.2 QoS Mappings (Generic Attributes)

User Level	*QoS Mapping (Application Level)*	Network Level (End to End)	*QoS Mapping (Network Level)*	Element Level (Hop to Hop)
Quality		Bandwidth boundary for the type of service (price)		Operating system resources (CPU)
Type of applications		Latency limit		Scheduling schemes
Cost		Jitter		Traffic flow weight
Availability		Loss/error rate		Queue length/delay
Guarantee (data)		Renegotiation/reconfiguration of SLA/QoS policy		Reconfigurable/programmable
Security				
Allow changes				
Maximum burst size				
Application packet size				
Maximum transfer unit (MTU)				

Netscript, Switchware, and spawning networks are based on the active network philosophy. The OPENSIG community argues that by using a set of open programmable network interfaces, open access to switches, routers, and base stations can be provided. Tempest and Xbind both use ATM as their underlying networking technology, to create virtual network architectures. Their approach separates network control from information transport and is mainly focused on programmable switches that provide some level of QoS support.

Active networks are based on dynamic deployment of new services at runtime, mainly within the boundaries of IP networks. Projects such as ANTS and smart packets provide dispatch, execution, and forwarding of packets based on the notion of active packets carrying the code and data and providing the combined control and data paths. These projects suggest the deployment of multiple coexisting execution environments through appropriate operating system support and active network encapsulation protocol.

The APIs of Xbind and ANTS are at two extremes, with regard to the flexibility and complexity of programming involved. Although the ANTS approach to the deployment of new APIs is highly dynamic, it presents a complex programming model in comparison to Xbind. The Xbind binding interfaces and programming paradigm are based on a set of CORBA and remote procedure call (RPC) mechanisms. In comparison to the capsule-based programmability used by ANTS, the Xbind approach is rather static in nature and the programming model is less complex. These two approaches show the two extremes of network programmability. An argument against the quasi-static API based on RPC (in the case of Xbind) could be that it is limited and restrictive. A counterargument is that the process of introducing and managing APIs is less complex than in the capsule-based programming paradigm. Similarly, it can be claimed that active message and capsule-based technologies are more open because of the inherent flexibility of their programming models, given that capsules can graft new APIs onto routers at runtime. Xbind and switchlets lack this dynamic nature, at the cost of simplified programming environments.

An important issue is the usage of Java by different active network projects. Because the standard JVM does not support access to transmission resources at a sufficiently low level, any implementation of ANTS or any other active network API based on Java cannot support QoS capabilities and is limited to the basic network capabilities provided by Java. However, at the University of Arizona, researchers have implemented a version of JVM called Joust,[38] which would support low-level abstraction for real-time scheduling.

The dynamic composition and deployment of new services can also be extended to include the composition of complete network architectures such as virtual networks. The Netscript project supports the creation of virtual network engines over IP networks. Virtual network engines interconnect sets of virtual nodes and virtual links to form virtual active networks. The Tempest framework supports the concept of virtual networks using safe partitioning over ATM hardware. The concept of physical switch partitions within the Tempest framework has led to the implementation of multiple coexisting control architectures. Tempest aims to address QoS through connection-oriented ATM technology and investigates physical resource sharing between alternative control architectures. Netscript supports the notion of sharing the underlying physical infrastructure in a customized way. Tempest and Netscript do not elaborate much on the security issues. The Switchware project elaborates quite extensively on security issues, but the required use of PLAN makes it inflexible. The spawning networks project is based on the Genesis Kernel, which is a virtual network operating system, to automate the creation, deployment, and management of distinct network architectures on the fly. Here, child virtual networks are created with their own transport, control, and management systems. A child network operates on a subset of its parents' network resources and in isolation from other spawned networks. At the lowest level of the framework, the transport environment delivers packets from source to destination end systems through a set of open programmable virtual router nodes called routelets. Routelets find a parallel with the virtual switches of Xbind and the switchlets of Tempest, which are implemented on ATM. However, routelets can operate over a wide variety of link-layer technologies, which is in contrast to switchlets, which are ATM based.

Analysis of active and programmable networks shows that projects such as Tempest, spawning networks, Darwin, VAN, and ANN all use static scheduling mechanisms. The bandwidth allocation to sessions in these projects can lead to unfairness and higher end-to-end delays due to the static nature of their packet scheduling. This problem of fairness and higher end-to-end delays becomes more prominent when there is congestion in the network. Furthermore, these projects do not use any congestion avoidance mechanism.

An integrated framework has been presented for adaptive open networks. Future networks are expected to be open systems where end users or third-party service providers can program or customize network elements to obtain the required services. The framework considers a number of key technologies that may be integrated to allow dynamic modification of the services offered over a network. The technologies employed include active and programmable networks, agent technologies, CORBA, and dynamically reconfigurable hardware. The framework also identifies different levels for the implementation of these technologies within the network. An active network manager has been proposed at the network level to take care of network operations, policy management, and security issues.

Although the higher layers provide adaptation of the service requirements, the QoS policy (e.g., RSVP versus DiffServ), and even the choice of network operator, it is only at the network element level that these changes are realized. Dynamically reconfigurable hardware (for example, runtime reconfigurable FPGAs) is used to effect the QoS target's changes on each router in a VPN. Such reconfigurable network elements are termed active routers. In operation, agents acting on behalf of the service providers are dispatched by the active network manager. Active packets from the manager are used to reconfigure the scheduling schemes within the active routers as new SLAs or QoS updates occur. It is expected that future network infrastructures will conform to a layered market model whereby sufficient alternatives exist that allow services to be traded as commodities. The agents facilitate these brokering activities for network resources and services in such an open heterogeneous network environment.

Exercises

1. Elaborate on the differences between traditional networks and programmable networks.
2. Discuss the binding architecture of Xbind.
3. How do capsules in ANTS transport codes between nodes? How is security implemented in ANTS?
4. What are the security features of Switchware?
5. Draw the packet format of a smart packet.
6. How do virtual network engines (VNEs) and virtual links (VLs) collaborate to form a Netscript system?
7. Discuss switchlets' creation process in Tempest. How is the control mechanism in Tempest implemented?
8. What are the major drawbacks of spawning networks?
9. Explain how the hierarchical fair service curve (H-FSC) scheduling mechanism works in Darwin.
10. Which component in Phoenix implements security for the framework?
11. Discuss the role of agents in the programmable framework.

References

1. D.L. Tennenhouse and D. Wetherall, Towards an Active Achitecture, paper presented at the Proceedings of Multimedia Computing and Networking (MMCN), San Jose, CA, January 1996.
2. J.M. Smith et al., Activating networks: a progress report, *Computer*, 32:4, 32–41, 1999.
3. D. Wetherall et al., Introducing new Internet services: why and how, *IEEE Network*, 12:3, 12–19, 1998.
4. Z. Fan and A. Mehaoua, Active Networking: A New Paradigm for Next Generation Networks?, paper presented at the Second IFIP/IEEE International Conference on Management of Multimedia Networks and Services (MMNS'98), France, November 1998.
5. D.L. Tennenhouse et al., A survey of active network research, *IEEE Commun. Mag.*, 35:1, 80–86, 1997.
6. A.T. Campbell et al., A survey of programmable networks, *ACM SIG-COMM Comp. Commun.*, 29:2, 7–23, 1999.
7. S.A. Hussain and A. Marshall, Provision of QoS using active scheduling, *IEE Proc. Commun.*, 151:3, 221–237, 2004 (special issue on Internet protocols, technology, and applications).
8. S.A. Hussain, An active scheduling paradigm for open adaptive network environments, *Int. J. Commun. Syst.*, 17:5, 491–506, 2004.
9. S.A. Hussain and A. Marshall, An agent based control mechanism for WFQ in IP networks, *Control Eng. Pract. J.*, 11:10, 1143–1151, 2003.
10. J. Biswas et al., IEEE P1520 standards initiative for programmable network interfaces, *IEEE Commun. Mag.*, 36:10, 64–70, 1998.
11. D.L. Tennenhouse et al., ANTS: A Toolkit for Building and Dynamically Deploying Networking Protocols, paper presented at IEEE OPENARCH'98, San Francisco, April 1998.
12. Mun Choon Chan et al., On Realizing a Broadband Kernel for Multimedia Networks, paper presented at the 3rd COST 237 Workshop on Multimedia Telecommunications and Applications, Barcelona, November 1996.
13. J.E. van der Merwe and I.M. Leslie, Service specific control architectures for ATM, *IEEE J. Selected Areas Commun.*, 16, 424–436, 1998.
14. J.E. van der Merwe et al., The Tempest: a practical framework for network programmability, *IEEE Network*, 12:3, 20–28, 1998.
15. S. Rooney, The Tempest: a frame for safe, resource-assured programmable networks, *IEEE Commun Mag.*, 36:10, 42–53, 1998.
16. I.M. Leslie, Shirking networking development time-scales: flexible virtual networks, *Electron. Commun. Eng. J.*, 11:3, 149–154, 1998.
17. A.T. Campbell et al., Spawning networks, *IEEE Network*, 13:4, 16–29, 1999.
18. A.T. Campbell et al., The Genesis Kernel: A Virtual Network Operating System for Spawning Network Architectures, paper presented at the Second International Conference on Open Architectures and Network Programming (OPENARCH), New York, 1999.

19. P. Chandra et al., Darwin: Customizable Resource Management for Value Added Network Services, paper presented at the Sixth IEEE International Conference on Network Protocols (ICNP'98), Austin, TX, October 1998.

20. P. Brunner and R. Stadler, Virtual Active Networks: Safe and Flexible Environments for Customer-Managed Services, paper presented at the 10th IFIP/IEEE International Workshop on Distributed Systems, Operations, Management, Zurich, October 1999.

21. G.M. Parulkar et al., A scalable high performance active node, *IEEE Network*, 13, 8–19, 1999.

22. L. Degrossi and D. Ferrari, A Virtual Network Service for Integrated-Services Internetworks, paper presented at the 7th International Workshop on Network and Operating System Support for Digital Audio and Video, St. Louis, 291–295, May 1997.

23. J.M. Smith et al., The Switchware active network architecture, *IEEE Network*, 12:3, 37–45, 1998.

24. J.M. Smith et al., A secure active network environment architecture: realization in SwitchWare, *IEEE Network*, 1998.

25. J. Touch and S. Hotz, The X-Bone, in *Third Global Internet Mini Conference in Conjunction with Globecom'98*, Sydney, November 1998, pp. 44–52.

26. B. Schwartz et al., Smart Packets for Active Networks, paper presented at the Second International Conference on Open Architectures and Network Programming (OPENARCH), New York, 1999, pp. 90–97.

27. Y. Yemini and S. Da Silva, Towards Programmable Networks, paper presented at the IFIP/IEEE International Workshop on Distributed Systems: Operations and Management, L'Aquila, Italy, October 1996.

28. S. Bhattacharjee, K. Calvert, Y. Chae, S. Merugu, M. Sanders, and E. Zegura, CANEs: An Execution Environment for Composable Services, Proceedings of the DARPA Active Networks Conference and Exposition (DANCE '02), IEEE, 2002.

29. D. Putzolu, S. Bakshi, S. Yadav, and R. Yavatkar, The Phoenix frame: a practical architecture for programmable networks, *IEEE Commun. Mag.*, 38:3, 160–165, 2000.

30. A. Kulkarni and G. Minden, Composing protocol frameworks for active wireless networks, *IEEE Commun. Mag.*, 38:3, 130–137, 2000.

31. A.A. Lazar et al., Realizing a foundation for programmability of ATM networks with the binding architecture, *IEEE J. Selected Areas Commun.*, 14:7, 1214–1227, 1996.

32. A.A. Lazar, Programming telecommunication networks, *IEEE Network*, 11:5, 8–18, 1997.

33. D. Wetherall, J. Guttag, and D. Tennenhouse, ANTS: Network services without the red tape, *IEEE Computer*, 32:4, 42–48, 1999.

34. S. Merugu, S. Bhattacharjee, E. Zegura, and K. Calvert, Bowman: A NodeOS for Active Networks, paper presented at IEEE INFOCOM, Tel Aviv, Israel, 2000, pp. 1127–1136.

35. E. Nygren, The Design and Implementation of a High Performance Active Network Node, MIT master's thesis, February 1998.

36. M. Shreedhar and G. Varghese, Efficient fair queuing using deficit round robin, *IEEE/ACM Trans. Networking*, 4:3, 375–385, 1996.
37. D. Chieng et al., A mobile agent brokering environment for the future open network marketplace, in *Seventh International Conference on Intelligence in Services and Networks (IS&N2000)*, Athens, February 2000, pp. 1-15.
38. J. Hartman et al., Joust: platform for liquid software, *IEEE Network*, 1998.

Chapter 5

Packet Scheduling for Active and Programmable Networks

5.1 Introduction

Data networks, including Internet Protocol (IP) networks, have long suffered from performance degradation in the presence of congestion. The rapid growth in both the use and size of computer networks has created a renewed interest in methods of congestion control for different classes of traffic. Congestion control has two areas of implementation. The first is at the source, where flow control algorithms vary the rate at which the source sends packets. Flow control algorithms are designed primarily to ensure the presence of free buffers at the destination host, but their role is more vital in limiting overall network traffic and in providing users with maximal utility from the network.

The second area of implementation is at the switches. Congestion can be controlled at the switches by assigning priority to services. Priority-based scheduling, if properly implemented, reduces congestion by controlling the order in which packets are sent and the usage of the switch's buffer space. It also determines the way in which packets from different sources interact with each other. Traditionally, the most commonly used scheduling packet algorithm is first come, first served (FCFS). In this scheme, the order of arrival of packets completely determines the bandwidth, promptness, and buffer space allocations, and it inextricably combines these three allocation issues. A source sending packets to a router at a sufficiently high speed can capture an arbitrarily high fraction of link bandwidth, starving delay-sensitive applications of bandwidth. Hence, FCFS cannot provide adequate congestion management in a typical heterogeneous network environment. Following the above line of reasoning, different scheduling algorithms that can discriminate among different classes of traffic have been proposed over the years. However, these scheduling mechanisms have many problems and drawbacks regarding the provision of quality of service (QoS) and fairness due to their static nature.

There has already been some research done within the area of reconfigurable Asynchronous Transfer Mode (ATM) switches.[1] The adaptive switch approach allows reconfiguration of the queuing scheme based on traffic profile, while maintaining a certain level of QoS for different classes of traffic. As a further step for the reconfiguration of scheduling mechanisms to modify the bandwidth allocation strategy of services in IP networks, a mechanism called *active scheduling* has been proposed.[2] Active scheduling allows the introduction of a procedure by which the queuing system of heavily loaded routers is dynamically altered according to user demands. This chapter describes in detail different scheduling schemes and the active scheduling scheme.

5.2 Packet-Scheduling Mechanisms

Packet scheduling is an important technique for system performance enhancement and QoS provisioning. Scheduling mechanisms can be broadly classified into work-conserving and non-work-conserving scheduling schemes. In a work-conserving discipline, a server is never idle when there is a packet to send. With a non-work-conserving discipline, each packet is assigned an eligibility time. A non-work-conserving server may be idle even if there are packets to be served. The well-known work-conserving algorithms include weighted-fair queuing (WFQ),[3] virtual clock (VC),[4] self-clocked fair queuing (SCFQ),[5] worst-case fair weighted-fair queuing (WF²Q),[6,7] and delay earliest-due-date (delay-EDD).[8] The non-work-conserving algorithms are jitter earliest-due-date (jitter-EDD),[8,9] stop-and-go,[10] and rate-controlled static priority.[11] Non-work-conserving mechanisms hold packets in regulators to reduce delay jitter and thus provide traffic shaping within the network.

To be useful in practice, a packet-scheduling algorithm should have the following desirable features:

Isolation of sessions: The algorithm must isolate an end-to-end session from undesirable effects of other misbehaving sessions. That is, the algorithm must be able to maintain the QoS guarantees for a session even in the presence of other misbehaving flows.

Low end-to-end delays: The algorithm must provide end-to-end delay guarantees for individual sessions. In particular, it is desirable that the end-to-end bound of a session depend only on the parameters of the session, such as its bandwidth reservation, and be independent of the behavior of other sessions. A higher end-to-end delay bound usually implies a higher level of burstiness at the output of the scheduler, and consequently requires larger buffers in the switches to avoid packet loss. Therefore, the delay bound affects not only the end-to-end behavior of the session, but also the amount of buffer space needed in the switches.

Utilization: The algorithm must utilize the link bandwidth efficiently.

Fairness: A scheduling scheme is called fair if the allocation of bandwidth to sessions is in proportion to the weights (allocated service rate) associated with the sessions. Unfair scheduling algorithms may offer most of the bandwidth only to a few sessions, leaving other sessions starved of bandwidth. Fairness is measured in terms of the worst-case fair index (WFI).

Implementation complexity: The scheduling algorithm must have a simple implementation. In packet networks with large packet sizes and lower speeds, a software implementation may be adequate to reduce the cost of implementation, but scheduling decisions must still be made at a rate close to the arrival rate of packets, if possible. For fast speeds, hardware implementation could be more appropriate, but the cost of implementation would be high. The implementation complexity of a scheduler is measured in terms of the complexity of the scheduling algorithm.

Scalability: The algorithm must perform well in switches with a large number of connections, as well as over a wide range of link speeds.

5.2.1 Weighted-Fair Queuing (WFQ)

WFQ is based on generalized processor sharing (GPS).[12] GPS is a general form of the head-of-line processor sharing (PS) service discipline. There is a separate FIFO queue for each session in PS sharing the same link. During any time interval when there are N nonempty queues, the server services the N packets at the head of the queues simultaneously, each at the rate of $1/N$ of the link speed. PS services all nonempty queues at the

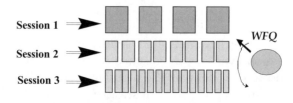

Figure 5.1 Weighted-fair queuing scheduling.

same rate; on the other hand, GPS allows different sessions to have different service shares and serves nonempty queues in proportion to the service share of their corresponding sessions. GPS is a hypothetical scheduling discipline that does not transmit packets as entities. It assumes that the server can serve multiple sessions simultaneously and that the traffic is infinitely divisible. GPS cannot be applied to the actual packet-based traffic scenarios, where only one session can receive service at a time and an entire unit of traffic, referred to as a packet, must be served before another unit is picked up for service. The simplest emulation of GPS is round-robin, which serves packets from each nonempty queue, instead of the infinitesimal amount that GPS would serve. There can be weights associated with the queues, and they receive service in proportion to these weights. Round-robin approximates GPS well when all connections have equal weights and all packets have the same size.[13] In case of different weights, weighted round-robin (WRR) serves a connection in proportion to its weight.[14] Weighted-fair queuing (WFQ), also called packet-by-packet generalized processor sharing (PGPS), approximates GPS when packets are of variable length,[12] as shown in Figure 5.1.

When a packet arrives at a WFQ scheduler, it is stamped with its virtual finish time. Virtual finish time is a service tag that indicates the relative order in which packets are to be served according to GPS and has nothing to do with the actual time at which packets are served. The server in WFQ serves the packets in the increasing order of their finish times. If the kth packet of session i arrives at time a^i_k and has length L^i_k, then virtual times at which this packet begins and completes service are given by S^i_k and F^i_k, respectively:

$$S^i_k = \max\left[F^{i-1}_k, V\left(a^i_k\right)\right]$$ (5.1)

$$F^i_k = S^i_k + \frac{L^i_k}{r_k}$$ (5.2)

where r_k is the minimum guaranteed service rate for session i and $V(t)$ is the virtual time function of the corresponding GPS system. If r is the link speed, then

$$r_k = \frac{\phi_i}{\displaystyle\sum_{j=1}^{N} \phi_j} \, r \qquad (5.3)$$

where ϕ_i is the relative weight of the session i and N is the number of sessions. Hierarchical WFQ (H-WFQ) is a hierarchical integration of single WFQ servers.[15] With the H-WFQ server, the queue at each internal node is a logical one, and the service it receives is distributed instantaneously to its child nodes in proportion to their relative service shares. This service distribution follows the hierarchy until it reaches the leaf nodes, where there are physical queues.

5.2.2 *Variants of Weighted-Fair Queuing (WFQ)*

Certain research groups have presented scheduling algorithms to address the problems associated with WFQ. These schemes are based on WFQ, but each scheme has certain drawbacks, as summarized in Table 5.1. The well-known variants of WFQ are WF²Q (worst-case fair WFQ), self-clocked fair queuing (SCFQ), and the virtual clock (VC) algorithm.

WF²Q[7] was developed to address the fairness problem of hierarchical WFQ, but WF²Q can have a higher worst-case fairness index (WFI) than WFQ under some circumstances. It has been proved by Fei and Marshall[16] that the use of WF²Q may cause the small-size packets to have a high WFI, compared to large packets in head of line (HOL) blocking. WFQ provides priority among the sessions, but hierarchical WFQ introduces a delay for real-time traffic, i.e., some packets receive more service than deserved. Consider the following example: if there are 11 sessions transmitting real-time traffic over a link of speed 1, session 1 has a guaranteed rate of 0.5, and the guaranteed rate for each of the other 10 sessions is 0.05. Session 1 sends 11 back-to-back packets starting at time 0, while all of the other sessions send one packet at time 0. If the server is GPS, then according to Equations 5.1 to 5.3, it will take 2 time units to transmit each of the first 10 packets of session 1, 1 time unit to transmit the 11th packet, and 20 time units to transmit the first packet from each of the other sessions.

Under WFQ, because the first 10 packets of session 1 have GPS finish times smaller than the packets of the other sessions, the server will service 10 packets of session 1 back to back before serving any packets from the

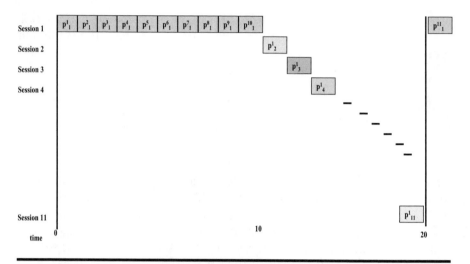

Figure 5.2 Service sequence of packets in WFQ.

other sessions. After the burst, the next packet on session 1, the 11th packet, will have a larger finishing time than the 10 packets at the head of the other sessions' queues; therefore, it will not be serviced until all of the other 10 packets are transmitted, and by that time, another 10 packets from session 1 will be serviced back to back, as shown in Figure 5.2. This can cause higher end-to-end delay for every 11th packet in a stream, resulting in higher delay jitter for sessions.

In WF²Q, rather than selecting the packet at the server from among all eligible packets from all queues, as in WFQ, the server considers only the set of packets that would have started receiving service in the corresponding GPS system at time t. It selects the packets that would complete service first in the corresponding GPS system (the packets may belong to the next session). Suppose in WF²Q at time 0 that all packets at the head of each session i queue, P_i^1, $i = 1 \ldots 11$, have started service, and among them, P_1^1 has the smallest finish time in GPS, so it will be served first in WF²Q. At time 1 there are still 11 packets at the head of the queues: P_1^2 and P_i^1, $i = 2 \ldots 11$.

Although P_1^2 has the smallest finish time, it will not start service in GPS until time 2; therefore, it would not be eligible for transmission at time 1. The other 10 packets from the other queues have all started service at time 0 at the GPS system and are thus eligible. Because they all finish at the same time in the GPS system, the rule of giving highest priority to the connection with the smallest number will yield P_2^1 as the next packet for service, as shown in Figure 5.3. At time 3, P_1^2 becomes eligible and has the smallest finish time among all packets; thus, it will be served next, and then at time 4, P_3^1 will be served.

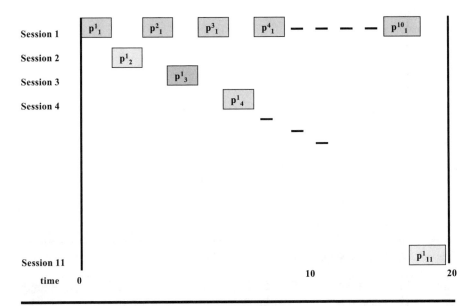

Figure 5.3 Service sequence of packets in WF²Q.

In self-clocked fair queuing (SCFQ),[5] when a packet arrives at an empty queue instead of using the system's virtual time to compute its (packet) virtual finish time, it uses the service tag or the virtual finish time of the packet currently in service. The packets are picked up for service in increasing order of associated tags. The virtual finish time of the arriving packets is computed by

$$F_k^i = \max\left[F_k^{i-1}, \hat{V}(a_k^i) \right] + \frac{L_k^i}{r_k} \qquad (5.4)$$

where $\hat{V}(t)$ is the virtual finish time of the packet currently being served, a_k^i is the arrival time of the packet, L_k^i is the size of the kth packet of session i, and r_k is the relative weight of the session. Because the virtual finish time of the packet currently in service is used, it can delay other packets of a session significantly, even though it sends packets according to the specified average rate.

In Figure 5.4 under SCFQ at time 0, P_1^1 has the smallest virtual finish time; therefore, it is served first. At time 1, all packets P_i^1, $i = 2 \dots 11$, have the virtual finish time of 20 under GPS, where i is the session. The first packet P_2^1 from session 2 will be served now. Because SCFQ uses the finish time of the packet currently in service as the current virtual

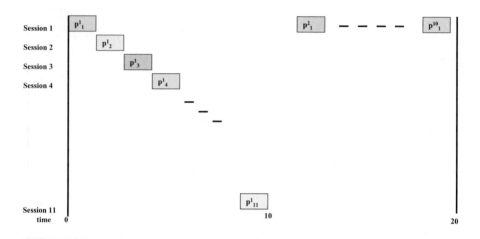

Figure 5.4   Service sequence of packets in SCFQ.

time value, $\hat{V}(1) = \hat{V}(2) = F_2^1 = 20$. As a result, when the second packet P_1^2 from session 1 arrives at time 2, its virtual finish time is set to be

$$F_1^2 = \max(F_{1,}^1 \hat{V}(2)) + \frac{L}{r_i} = \max(2, 20) + 2 = 22$$

where $L = 1$ and $r_i = 0.5$ are the size and guaranteed rate of P_1^2, respectively. Hence, among all the packets ready to be served, P_1^2 has the largest finish time. It will not start service until all of the other 10 packets P_i^1, $i = 2 \ldots$ 11, are serviced. This incurs a higher delay for the packets of session 1.

A similar approach was proposed under the name start-time fair queuing,[17] where the starting time of the packet currently in service is used to compute the time stamp of the arriving packet. The packets are then serviced in the increasing order of the start tags. This approach reduces the complexity of the algorithm, but the price is paid in terms of higher end-to-end delay bounds that grow linearly with the number of sessions sharing the outgoing link.

Virtual clock[4] discipline aims to emulate a time division multiplexing (TDM) system. Each packet is allocated a virtual transmission time, which is the time at which the packet would have been transmitted were the server actually performing TDM. Packets are transmitted in increasing order of virtual transmission times. Each data flow is assigned a virtual clock that ticks at every packet arrival from that flow; the tick step is equal to the mean interpacket gap. In this way, the virtual clock reading tells the expected arrival time of the packet. If a flow sends packets according to its specified average rate, its virtual clock reading should be in the vicinity of the real-time. To imitate the transmission ordering of a TDM system, each switch may stamp packets by the flows' virtual clock values and use the stamps to order transmissions, as if the virtual clock stamp were the real-time slot number in the TDM system. One major difference between a virtual clock-controlled packet-switching network and a TDM system is that the virtual clock algorithm merely orders packet transmission without changing the statistical sharing nature of packet switching; the network forwards all packets as long as resources are available. Another major difference is that packet networks can support arbitrary throughput rates of individual flows. The network reservation control determines how much share of the resources each flow may take on average; if more than one packet is waiting, the virtual clock algorithm determines which packet should go next based on the flows' reserved transmission rates. The virtual clock also plays the role of a flow meter driven by packet arrivals. Because it is advanced according to the flow's specified average transmission rate, the difference between the virtual clock and the real-time clock shows how closely a running flow is following its claimed rate. A feedback to flow sources can adjust the throughput. The virtual clock algorithm provides the same end-to-end delay and burstiness bounds as WFQ, but with a simple time-stamp computation algorithm; however, it is unfair over a period of time. A backlogged session in the VC server can be starved for an arbitrary period as a result of excess bandwidth it received from the server when other sessions were idle.

5.2.3 Non-Work-Conserving Algorithms

In a work-conserving algorithm, a server is never idle whenever there is a packet to send, whereas in a non-work-conserving algorithm, the server may be idle if there are packets to serve. Non-work-conserving algorithms hold packets in regulators (buffers), and thus each packet is assigned an eligibility time. Well-known non-work-conserving algorithms follow.

5.2.3.1 Earliest-Due-Date Schemes

The earliest-due-date (EDD)[8] scheduler assigns each packet a deadline, and the scheduler serves packets in order of their deadline. With EDD, the packets assigned deadlines closer to their arrival times receive a lower delay than packets assigned deadlines farther away from their arrival times. Delay-EDD[8] is an extension of EDD that specifies the process by which the scheduler assigns deadlines to packets. The deadline of a packet is the sum of the local delay bound d and the expected arrival time of the packet. The service discipline and the admission control policy ensure that the packet is guaranteed to leave before the deadline or, at most, d time units after the expected arrival time of the packet. It is possible that a packet is delayed longer in a server than its local delay bound. However, this happens if a packet's expected arrival time is larger than its actual arrival time, which means that the packet is ahead of schedule in previous servers. The jitter-EDD[9] discipline extends delay-EDD to provide delay jitter bounds. After a packet has been served at each server, a field in its header is stamped with the difference between its deadline and the actual finishing time. A regulator at the entrance of the next server holds the packet for this period before it is eligible to be scheduled. Jitter-EDD can be considered a combination of an earliest-due-date scheduler and regulators. If $d_{i,k}$ is the delay bound assigned to channel i at the kth node along its path, the deadline dl_n, assigned in node n to a channel i packet, that entered the network at time t_0 is

$$dl_n = t_0 + \sum_{k=1}^{n} d_{i,k} + P_n \tag{5.5}$$

where P_n is the propagation delay from the source to node n. As a result of the above equation, the jitter of packets on a real-time channel at its exit point from the network equals the jitter introduced by the last node in the network. This reduces the delay jitter but increases the end-to-end delay of sessions. Jitter-EDD assigns the bandwidth at the peak rate with no statistical multiplexing.

5.2.3.2 Stop-and-Go

Stop-and-go uses a framing strategy.[10] In this scheme, the time axis is divided into frames, which are periods of some length T. Packets are transmitted in frames. Stop-and-go defines departing and arriving frames for each link. At each switch, the arriving frame of each incoming link is mapped to the departing frame of the output link by introducing a constant delay, say α, where $0 \leq \alpha \leq T$. In this scheme, the transmission of a packet that has arrived on any link during a frame should always be postponed until the beginning of the next frame. Because packets arriving during a frame of the output link are not eligible for transmission until the next frame, the output link may be left idle even when there are packets to be transmitted. Stop-and-go ensures that packets on the same frame at the source stay in the same frame throughout the network. The framing strategy introduces the problem of coupling between delay bound and bandwidth allocation granularity. The delay of any packet at a single switch is bounded by two frame times. To reduce the delay, a smaller T is desired. Because T is also used to specify traffic, it is related to bandwidth allocation granularity. If the size of a packet is P, the minimum granularity of bandwidth allocation is P/T. To have more flexibility in bandwidth allocation or a finer bandwidth allocation granularity, a larger T is required. So, a low delay bound and fine granularity of bandwidth allocation cannot be achieved simultaneously in stop-and-go. To implement a stop-and-go scheme, mechanisms are needed at two levels: the link level and the queue management level. At the link level, a frame structure is needed and there is a synchronization requirement such that the framing structure is the same at both the sending and receiving ends of the link. At the queue management level, two FIFO queues are needed for each priority level, one for storing the eligible packets ready to be transmitted and the other for storing packets that would not be eligible until the end of the current frame time. Mechanisms are needed to swap the two FIFO queues at the start of each frame time. Also, the set of FIFO queues with eligible packets needs to be serviced according to a static priority policy.

5.2.3.3 Rate-Controlled Static Priority (RCSP)

The goal of RCSP[11] is to achieve flexibility in the allocation of delay and bandwidth as well as simplicity of implementation. RCSP has two components: a rate controller and static priority scheduler. The rate controller consists of a set of regulators corresponding to each of the connections traversing the server; each regulator is responsible for shaping the input traffic of the corresponding connection. Upon arrival of a packet, an eligibility time is calculated and assigned to the packet by the regulator. The packet is held in the regulator until its eligibility time, before being handed to the scheduler for transmission. RCSP can use rate jitter and delay jitter regulators, as shown in Figure 5.5.

The scheduler in RCSP uses a static priority (SP) policy, i.e., it always selects the packet at the head of the highest-priority queue that is not empty. The SP scheduler has different numbers of priority levels, each corresponding to a different delay bound. Each connection is assigned to a priority level during connection establishment. Multiple connections can be assigned to the same priority level, and all packets on the connections associated with a priority level are queued to the end of the queue for that priority level. In an RCSP server, there is a regulator for each connection and the regulated traffic on each connection can be assigned to any priority level in the scheduler. In the case of stop-and-go, server regulators are associated with priority levels in the scheduler. There is a one-to-one correspondence between the regulator and the priority level.

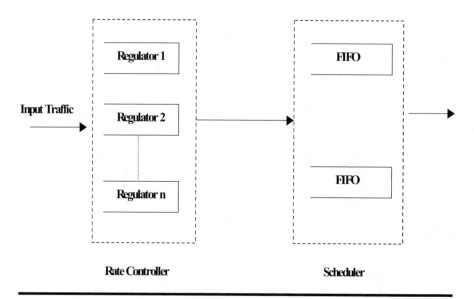

Figure 5.5   Rate-controlled scheduler.

The traffic on a connection has to be specified with respect to the frame size, which is the same as the connection's local delay bound. Table 5.1 gives a summary of all the scheduling mechanisms described in Section 5.2. It presents a description and drawbacks of each scheme.

5.2.4 Analysis of End-to-End Delay and Delay Jitter (Delay Variation) Characteristics

Table 5.2 presents the end-to-end characteristics for different scheduling mechanisms. If a connection can satisfy the traffic constraints and is allocated the required buffer space, it can be guaranteed an end-to-end delay bound and delay jitter bound as listed in Table 5.2. In the table, C_i is the link speed of the ith switch on the path traversed by the connection, k_i is the number of connections sharing the link with the connection at the ith switch, r_j is the guaranteed rate for the connection, L_{max} is the largest packet size, and σ_j is the maximum burst size. $D(b_j, b^*)$ is the worst-case delay. This is a function of the traffic constraint $b^*(.)$ used in the regulators and traffic constraint $b_i(.)$ used to specify the source. Traffic constraint is the number of bits allowed on a connection.

In Table 5.2,[4,12] it may be observed that even though virtual clock, WFQ, and WF²Q have some differences, they provide identical end-to-end delay bounds for connections. If the delay bound provided by these schedulers and that provided by the ideal GPS discipline are compared, it is seen that they share the first term, σ_j/ρ_j, which is the time needed to send a burst of size σ_j in a GPS with a guaranteed rate of r_j. The other terms in virtual clock, WFQ, and WF²Q show that the traffic is not infinitely divisible and the server needs to serve one packet at a time. The table also shows that if the guaranteed rate is the same, then the delay bound provided by SCFQ is larger than that provided by virtual clock, WFQ, and WF²Q. This is due to the inaccuracy introduced by the approximation algorithm in calculating the virtual time.

For all four disciplines, because the server allocates service shares to connections in proportion to their average rates, there is a coupling between the end-to-end delay bound and bandwidth provided to each connection. In particular, the end-to-end delay bound is inversely proportional to the allocated long-term average service rate. Thus, in order for a connection to get a low delay bound, a high-bandwidth channel needs to be allocated. This results in a waste of bandwidth when the low delay connection also has low throughput.

5.2.5 Complexity Analysis of Work-Conserving Algorithms

The complexity of an algorithm is the estimate of the amount of time needed to solve a problem by using that algorithm. The complexity is measured in terms of the number of operations required by the algorithm instead of the actual CPU time the algorithm requires. The efficiency of algorithms can be compared in terms of number of operations. For example, if N is the number of inputs to an algorithm, algorithm X requires a number of operations proportional to N^2 and algorithm Y requires a number of operations proportional to N. If N is 4, then algorithm Y will require 4 operations, while algorithm X will require 16 operations. Hence, algorithm Y is more efficient than X, because Y requires time proportional to $f(N)$. Algorithm Y is said to be of order $f(N)$, denoted by $O(f(N))$, where $f(N)$ is called the algorithm's growth rate function.

For packet schedulers, complexity has two domains: (1) complexity of computing the system virtual time function and (2) complexity of sorting the time stamps. The system virtual time function, $V(t)$, is the normalized fair amount of service that all backlogged sessions should receive by a certain time in the GPS system. $V(t)$ is a piecewise linear function; its slope at any point of time t is inversely proportional to the sum of the service shares of sessions in the set of backlogged sessions. Whenever a session becomes backlogged or served in the corresponding GPS system, the slope of $V(t)$ changes. Therefore, evaluation or updating of $V(t)$ requires keeping track of the set of backlogged sessions and its evolution in time. The computational complexity associated with the evaluation of $V(t)$ depends on the frequency of breakpoints in $V(t)$, i.e., the frequency

Table 5.1   Comparison of IP Packet-Scheduling Schemes

Scheduling Scheme	Description	Drawbacks
First come, first served (FCFS)	Packets are served in the order they arrive.	No protection between well-behaved and misbehaved connections; cannot provide QoS guarantees.[18]
Weighted round-robin (WRR)	Packets are served in round-robin fashion in proportion to the weight of the connection.	Bandwidth allocation problem if the server cannot predict the source's mean packet size in advance; it is fair over timescales longer than a round time; at shorter timescales, some connections may get more service than others; if a connection has a small weight or the number of connections is large, this may cause longer periods of unfairness.[19]
Weighted-fair queuing (WFQ)	Server tags packets with finishing times under GPS, then serves packets in order of tags.	Allocation of weights is essentially static in a scheduler or across a number of schedulers over a period; delay and bandwidth coupling problem; there is no provision of intelligent load management over a set of routers spanning the VPN.[20]
Worst-case fair weighted-fair queuing (WF²Q)	Similar to WFQ, with the difference that the server chooses the packet with the smallest finish time among all the packets that have already started service in the corresponding GPS emulation.	Same as WFQ; has higher worst-case fairness index (WFI) under some conditions than WFQ; additionally, WF²Q has the same virtual time calculation complexity as WFQ.[16]

Scheme		
Virtual clock (VC)	Emulates time division multiplexing; server tags packets with completion times under TDM, then serves packets in order of tags.	VC has the same latency as WFQ, but is not a fair algorithm; needs per connection state; packets have to be at least partially sorted before being served.[3]
Self-clocked fair queuing (SCFQ)	Same as WFQ, except that round number is set to finish number of packet in service.	Same as WFQ; worst-case end-to-end delay is much larger than WFQ, because latency is a function of the number of sessions that share the outgoing link.[21]
Delay earliest-due-date (delay-EDD)	Server tags packets with a deadline that is the sum of the expected arrival time and the local delay bound; packets are served in order of their deadline.	Bandwidth must be reserved at the peak rate instead of at the average rate; needs per connection state and sorted priority queue; more buffers are needed.[9]
Jitter earliest-due-date (jitter-EDD)	Consists of a regulator and scheduler; regulator reconstructs traffic to original form.	Similar to delay-EDD; separate regulator makes implementation complex; to reduce delay jitter, all packets receive a large delay.[18]
Stop-and-go	Packets are transmitted in frames; maps the arriving frame of each incoming link to the departing frame of the output link by a constant delay.	Introduces the problem of coupling between delay bound and bandwidth allocation.[18]
Rate-controlled schemes (e.g., RCSP)	Packets are assigned eligibility time before being handed to the scheduler; can use either rate jitter or delay jitter regulators.	With delay jitter regulation, sources must be policed; separate regulator complicates implementation; to satisfy delay jitter bound, all packets receive a large delay.[18]

Table 5.2 End-to-End Delay and Delay Jitter

Scheduling Scheme	End-to-End Delay Bound	Delay Jitter Bound
GPS	σ_j/r_j	σ_j/r_j
Virtual clock	$\dfrac{\sigma_j + nL_{\max}}{r_j} + \displaystyle\sum_{i=1}^{n} \dfrac{L_{\max}}{C_i}$	$\dfrac{\sigma_j + nL_{\max}}{r_j}$
WFQ and WF²Q	$\dfrac{\sigma_j + nL_{\max}}{r_j} + \displaystyle\sum_{i=1}^{n} \dfrac{L_{\max}}{C_i}$	$\dfrac{\sigma_j + nL_{\max}}{r_j}$
SCFQ	$\dfrac{\sigma_j + nL_{\max}}{r_j} + \displaystyle\sum_{i=1}^{n} k_i \dfrac{L_{\max}}{C_i}$	$\dfrac{\sigma_j + nL_{\max}}{r_j} + \displaystyle\sum_{i=1}^{n} (k_i - 1) \dfrac{L_{\max}}{C_i}$
Stop-and-go	$nT_j + \displaystyle\sum_{i=1}^{n} \alpha_i$	T_j
Jitter-EDD	$D(b_j, b^*) + \displaystyle\sum_{i=1}^{n} d_{i,j}$	$D(b_j, b^*) + d_n$

of transitions in and out of the set of backlogged sessions. Such transitions can be rather infrequent on average, and sometimes a number of them can occur during a single-packet transmission time. The reason for this behavior is that in the GPS system, where $V(t)$ is to be determined, packets are not served one after another; instead, one packet from every back-logged session is in service simultaneously. It is possible that many packets may finish service not exactly at the same time. With each packet finishing service, the corresponding session could become absent, if there are no other packets from that session, thereby creating a change in $V(t)$. If there are N sessions that can be backlogged or served, during an arbitrarily small interval, the worst-case complexity in this case will be $O(N)$. Table 5.3 lists virtual time computation complexities of different scheduling algorithms.

The second type of implementation complexity is the complexity of sorting the time stamps of packets. Each packet has a virtual start time and a virtual finish time. These are the times a packet starts and finishes service in the GPS system. The virtual finish time can also be interpreted

Table 5.3 Virtual Time Complexities of Work-Conserving Schedulers

Scheduling Scheme	Complexity of System Virtual Time
WFQ	$O(N)$
WF²Q	$O(N)$
SCFQ	$O(\log N)$
STFQ	$O(\log N)$
Virtual clock	$O(\log N)$

as the amount of service normalized with respect to the service share a session has received right after another packet of this session is served. Packets in the work-conserving schedulers are sorted out for service according to their start times or finish times. The sorting complexity for these schedulers can be $O(N)$ or $O(\log N)$. $O(N)$ represents a linear search, and $O(\log N)$ represents a binary search for sorting, which is computationally less expensive.

In WFQ, a maximum of N events may be triggered during the transmission of one packet, where N is the number of sessions that share the outgoing link and the packets are served according to the increasing order of the virtual finish times, or smallest virtual finish time first (SFF), from all the sessions. Thus, the system virtual time complexity of WFQ is of the order of $O(N)$. The WF²Q server considers only the set of packets that have started receiving service in the corresponding GPS system at time t and selects the packet among them that has the smallest eligible virtual finish time, rather than selecting the packet from among all the packets at the server, as in WFQ. This improves the fairness in WF²Q, but because packets are selected from among N sets of sessions or queues, WF²Q has the system virtual time complexity of $O(N)$. Thus, the virtual time complexities of WFQ and WF²Q increase linearly with the number of sessions. SCFQ does not use GPS virtual time; instead, it uses the virtual finish time of the packet currently in service. Hence, the virtual time computation complexity is $O(\log N)$. Similarly in start-time fair queuing (STFQ), virtual time is calculated by using the start tag of the packet in service. The computation of virtual time in STFQ is less complex because it only involves examining the start time of the packet in service. Its complexity is $O(\log N)$.

5.2.6 *Fairness Analysis of Work-Conserving Algorithms*

The fairness issues have been discussed in detail in Bennett and Zhang.[6] They define a metric called worst-case fair index (WFI). WFI quantifies the service discrepancy between a packet discipline and GPS. According to the definition of WFI given in Bennett and Zhang,[6] a service discipline is called worst-case fair for session i if for any time T, the delay of a packet arriving at T is bounded above by $\dfrac{Q_{i,s}(T)}{r_i} + C_{i,s}$, i.e.,

$$d^k_{i,s} < a^k_i + \frac{Q_{i,s}(a^k_i)}{r_i} + C_{i,s} \tag{5.6}$$

where r_i is the throughput guarantee to session i, $Q_{i,s}(a^k_i)$ is the queue size of session i at time a^k_i, and $C_{i,s}$ is the worst-case fair index for session i at server s. The worst-case index of GPS is always zero. The normalized worst-case index is defined as

$$c_s = \max_i \{c_{i,s}\} \tag{5.7}$$

where $c_{i,s} = r_i \dfrac{C_{i,s}}{r}$. Intuitively, $C_{i,s}$ is the maximum time a packet coming to an empty queue needs to wait before starting to receive its guaranteed service. A service discipline is called worst-case fair if it is worst-case fair for all sessions. Smaller values of WFI are always desirable to have less queuing delay and better fairness. In hierarchical WFQ,[6] if link bandwidth is shared by a best-effort and real-time session, and the real-time traffic is not sent for a time period, then the total bandwidth would be consumed by the best-effort traffic for the current time period. When the real-time traffic is sent, the total available bandwidth will be in deficit; therefore, it has to wait for a long time period, during which traffic from other sessions can catch up with their service shares. This leads to unfairness for the real-time session and higher delays. Such problems do not exist for GPS, as no session can get more bandwidth than its fair share.

5.3 Active Scheduling for Programmable Routers

Active scheduling allows the introduction of a procedure by which the functionality of routers can be dynamically modified.[24] Such adaptation may be initiated by user applications, third-party service providers, or a network operator. In this scheme, the queue weights of heavily loaded routers spanning the virtual private network (VPN) are dynamically altered according to different classes of traffic (voice, video, and data), because each class has different traffic characteristics. Dynamic adaptation of weights according to either traffic load or user demand offers the potential to implement prioritized fair queuing and service class guarantees based on traffic load profile and negotiated service parameters, such as end-to-end delay, delay variation, and price. This is achieved by using agents on the routers initiating the reconfiguration of weights.

In the context of active scheduling, agents are regarded as software programs designed to carry out the function or task of changing the weights of sessions dynamically on behalf of an operator by communicating with other agents. Each VPN in an active router environment contains two types of agents on the active routers: queue agents and control agents. *Queue agents* control the delay of the services according to the maximum and minimum delay bounds specified by the clients. Whenever there is an increase in the queuing delay of the session, the agent associated with the queue increases the weight of that queue, and decreases its weight when there is a decrease in the queuing delay. *Control agents* control the end-to-end delay of a VPN by monitoring the weight of a session on a downstream router. Whenever there is an increase in the weight of a session in the downstream router, it informs the upstream router to reduce the weight of the session to a certain limit, so that the traffic load that has increased the queuing delay of the loaded downstream router is quickly cleared and the end-to-end delay for sessions is maintained within the bounds.

Active scheduling is functionally compatible with the DiffServ architecture. Agents can be used to monitor the queuing delay of aggregates in DiffServ. If more fine-grained control is required, agents can be used on a per flow basis, or one agent for N number of sessions within aggregates if the QoS requirements for the N set of sessions are the same. The weights will be changed whenever there is congestion on the links or a change in the number of flows. The number of agents on a router depends on the aggregates or classes of service. The control agents activate or trigger the control agent in the upstream router in the immediate neighborhood by sending a lightweight weight-change signal (trigger) to it. Active scheduling is scalable because the weight change is always between two immediate routers (upstream and downstream), instead of throughout the routers spanning the VPN. The downstream router sends the weight-change signal to the upstream router by reserving a path between the two routers. This is accomplished by using a high-priority IP packet over that path.

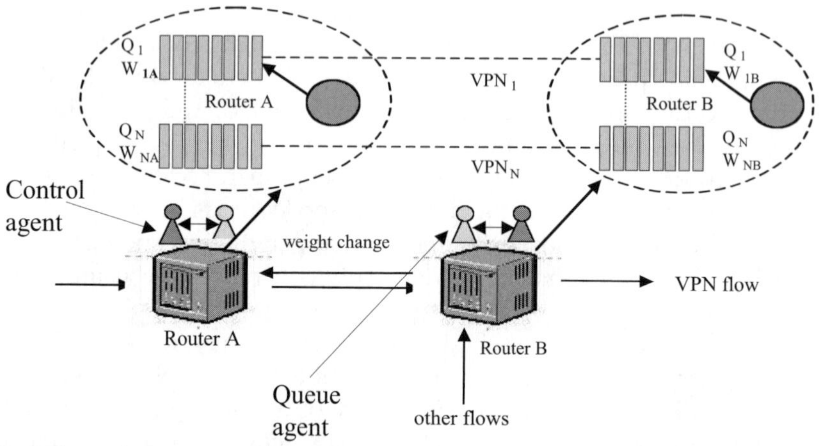

Figure 5.6 Weight adjustments by agents in routers.[20] (Copyright © 2004, John Wiley & Sons Ltd. Reproduced with permission.)

In Figure 5.6, routers A and B are active. Q_1 in routers A and B forms the VPN for session 1. Similarly, other sessions in the network form the VPNs through the queues of active routers A and B. Q_1 of active router B is loaded by active router A and other routers in the network. When the delay of queue Q_1 of router B reaches its maximum limit, a control agent on router B informs the *control agent* on router A to reduce the weight W_{1A} of its queue Q_1 to a lower bound (communicated to the control agent by the queue agent) by sending a weight-change signal. This reduces the traffic load on Q_1 of active router B, and its queuing delay decreases. When the weight of Q_1 in B decreases, the weight of Q_1 in router A is again increased. Thus, the queuing delay of Q_1 in active router B is maintained between the limits by increasing and decreasing the weights of Q_1 between the limits in active router A. This technique provides fairness among the sessions and keeps the end-to-end delay for Q_1 on routers A and B within bounds by spreading the bursts over the queues of upstream routers. Figure 5.7 shows the weight-change profile of Q_1 in router A.

5.3.1 Motivation for Active Scheduling

A virtual private network (VPN) can be provided by a third-party service provider (or network operator) over an existing network to offer dedicated connectivity to a closed group of users. VPNs are more economic to install and are created in a shorter timescale than deploying and maintaining physical cables and equipment themselves. However, in current IP networks employing VPNs, there is a trade-off between simplicity and QoS.

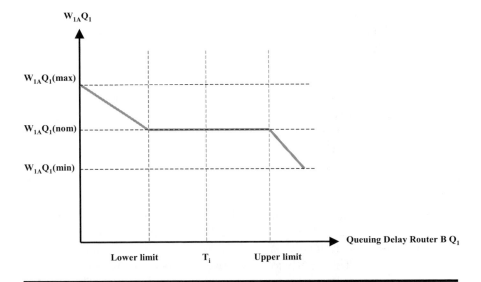

Figure 5.7 Weight-change profile of active router A and delay in router B.

Resource-guaranteed VPNs can easily be created, but they use static scheduling schemes for bandwidth allocation. One of these well-known static scheduling schemes is WFQ. However, there are certain problems with WFQ.

First, the delay bounds in WFQ are inversely proportional to the minimum guaranteed service rate and also to the long-term sustainable rate due to the rate-proportional weighting of sessions. That is, if there is an increase in the throughput of a session or if there is a burst, the weight will be increased to reduce the delay. The other sessions, which are sharing the same link, shall be starved of bandwidth and their delays increase. This is called *coupling of delay and bandwidth.*

Second, it allocates a fixed amount of bandwidth over a defined timescale to the sessions over the routers spanning the VPN; that is, the assignment of weights to any queue within a scheduler, or across a number of schedulers, is essentially static. This leads to bandwidth bottlenecks at the output link of routers for the sessions joining the queues of the router from other sections of the network.

Third, there is no provision for *intelligent load management* or *intelligent load control* in static WFQ. This results in higher end-to-end delays for the services and waste of bandwidth. This is due to the fact that different routers in the network are loaded with different traffic loads and different services have different QoS requirements. For example, if a router in the core network spanning the VPN is heavily loaded and the upstream routers are lightly loaded, or the upstream routers are loaded with less time critical services, then this can lead to waste of bandwidth over the links and higher end-to-end delays for time-critical services.

Active scheduling addresses the problems of WFQ by dynamically adjusting the weights of the sessions according to the negotiated delay bounds of the clients and changing network conditions (burstiness, change in the number of sessions, etc.) on a per node basis over a period. It uses *intelligent load control* for bounded end-to-end delays. When the queuing delay of a loading session reaches its maximum value, the agent monitoring the queue informs the control agent on the upstream router to decrease its weight. This reduces the service rate of the loading session and keeps the end-to-end delay of all services within bounds. Active scheduling maintains the delay bounds of a session by defining the upper and lower weight limits on a per node basis for that session. The weight of a session does not fall below a nominal level; it increases to a maximum value when the delay increases and decreases to a lower value when the delay decreases. The bandwidth left over by decreasing the weight is shared among the sessions according to their weight limits. This mechanism provides a simple and accurate means of serving the sessions.

5.3.2 Mathematical Model and Algorithm of Active Scheduling

The algorithm for agent-based WFQ keeps the end-to-end delay of each session's VPN within bounds by using agents. In this scheme, an agent in the loaded router informs the agent in the upstream router to reduce its weight so that the traffic load that increases the queuing delay of the loaded router is cleared quickly and the end-to-end delay for the sessions is maintained within bounds.

According to this approach, recalibration of weights is performed periodically after a defined period. A range or set of limits of queuing delay is defined for each session. This range is based on the lower and upper limits of queuing delays, as shown in Figure 5.8.

The reconfiguration of weights for a session i is based on the delay calculated for the session i. The delay tolerance is given by

$$T_{tol} = T_{i\max} - T_{i\min} \tag{5.8}$$

If the measured delay T_i falls outside the limits, the weights for the session are recalibrated to maintain

$$T_{i\max} \leq T_i \leq T_{i\min} \quad \forall I \in B(t) \tag{5.9}$$

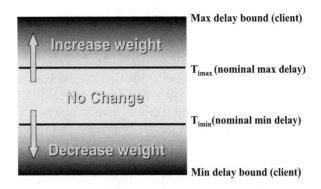

Figure 5.8   Weight change in an active router.[20] (Copyright © 2004, John Wiley & Sons Ltd. Reproduced with permission.)

where $B(t)$ is the set of sessions currently backlogged. T_{imax} is the nominal maximum delay that a session can have. This delay bound is always less than the maximum delay limit set for the client by the network operator. As the queuing delay reaches T_{imax}, the weight for the session is increased by the agent monitoring the queuing delay. T_{imin} is the nominal minimum acceptable delay of the session. When T_i is at or less than T_{imin}, the weight for the session is decreased by the agent, and the excess bandwidth is shared among the other sessions according to their weight ratios. In some cases, T_{imin} can be zero (no delay), i.e., no packets in the queues. There is no change in weights when T_i is between T_{imax} and T_{imin}. Preference is given to the delay-sensitive services such as Motion Picture Experts Group (MPEG) 2 video and Voice-over-IP (VOIP) for weight allocations. If all the bandwidth has been consumed, no new session is served.

Whenever a packet joins the queue of a session *i*, it is enqueued in the queue. The packet waits in the queue until it is served by the scheduler. The queuing delay of the packet is calculated by the following equations:

$$sum\_delay + delay\_arrival\ (i)$$

$$T_i = sum\_delay/delay\_arrival\_index$$

where *sum_delay* is the accumulated delay, *delay_arrival* is the delay of each packet, and T_i is the measured delay calculated per second.

If weight (queue) < nominal weight + upper_ weight limit and if the queuing delay is more than the upper limit, then the weight of the queue is increased by $\Delta(w+)$, i.e.,

$$w_i(t + \tau) = w_i(t) + \Delta(w+) \text{ if } T_i \geq T_{imax} \qquad (5.10)$$

This weight increase dequeues the packet and brings down the *sum_delay* and, correspondingly, T_i comes down to T_{imin}.

When the weight of the queue in a heavily loaded router reaches its maximum limit, the agent monitoring the queue informs the upstream router to reduce the weight of the same queue in the upstream router, by sending a weight-change signal to the *control agent* on the upstream router, i.e.,

If weight (queue) = upper_weight limit, then send weight-change signal to the control agent on the upstream router.

When the weight of the loaded queue comes down to the nominal weight, the weight of the queue of the upstream router is again increased.

If weight (queue) > nominal weight + lower_ weight limit and if the queuing delay is less than the lower limit, then the weight of the queue is decreased by $\Delta(w-)$, i.e.,

$$w_i(t + \tau) = w_i(t) - \Delta(w-) \text{ if } T_i \leq T_{imin}, T_i \neq 0 \qquad (5.11)$$

If the delay of the queue remains between the max and min limits, then the weight of the queue remains at the nominal value, i.e.,

$$w_i(t + \tau) = w_i(t) \text{ if } T_{imin} \leq T_i \leq T_{imax} \qquad (5.12)$$

where $w_i(t)$ is the nominal weight and $\Delta(w+)$ and $\Delta(w-)$ are the weight increment and decrement values whenever there is an increase or decrease in the queuing delay of a session. The Δw values depend on the type of weight-change profile used; e.g., the piecewise linear weight change increases the weight in unit steps, whereas the step function increases the weight to the maximum limit as a single step change.

This mechanism reduces any short-term unfairness by dynamically changing the weights within the bounds. That is, if there is a bursty traffic source, its traffic is controlled by reducing its weight in the upstream routers. This avoids any short-term unfairness among the services due to bursty sources. If D_{imax} is the end-to-end delay for a queue i,

$$D_{i\,max} = \sum_{i=1}^{N} T_{i\,max} \qquad (5.13)$$

where N is the number of routers spanning the VPN. Because the end-to-end delay depends on the reserved bandwidth, it is easy to reserve bandwidth along the path of a VPN to meet the QoS requirements. As a result, the flow may never have high backlogged traffic in each router. The worst-case queuing delay bound of a session when there is high congestion on the link is calculated by the equation

$$T_{i\,\mathrm{max}} = \frac{burstsize}{r_i} + \frac{Packetsize}{r_i} + \frac{MTU}{r} \tag{5.14}$$

where *burstsize* and *Packetsize* are an application's burst size and packet size, respectively, and r_i is the bandwidth allocated to the application according to Equation 5.3. The weights of the application vary between maximum, minimum, and nominal bounds. r is the link rate and *MTU* is the maximum transfer unit (MTU) of traffic on the link. The values of burst size, Packet size, and MTU are service level agreement (SLA) parameters. For example, the burst size of MPEG2 depends on the size of the I frame, and the burst size and packet size of voice depend on the coding rate and number of frames per packet. Figure 5.9 shows the flowchart for weight change in active routers spanning a VPN.

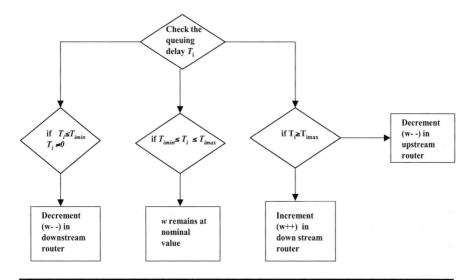

Figure 5.9 Flowchart of weight change in active routers.

5.4 Summary

Congestion management and congestion avoidance are important aspects of QoS assurance when there are high congestion in the network and high demand of bandwidth from the services. Because traditionally Internet traffic is treated as a best effort, the routers forward the packet to the next hop as it arrives, without any notion of priority to the services. This leads to higher end-to-end delay, packet loss, and jitter. Thus, a QoS-aware packet scheduler with a proper congestion avoidance mechanism provides better QoS by using priority among different classes of service.

Packet schedulers fall into two categories: work-conserving and non-work-conserving. With the exception of delay-EDD, all working-conserving algorithms described are based on the GPS approach, which serves an infinitesimally small amount of data from queues. These schemes emulate the idealized GPS approach by assigning a virtual time to the packets and serve the packets according to their virtual times. The scheduling mechanisms are analyzed on the basis of end-to-end delay and delay jitter. Different scheduling mechanisms within the same category offer different end-to-end delay and delay jitter bounds, e.g., SCFQ offers higher end-to-end delay than WFQ and WF^{2}Q.

To provide different qualities of service to different clients, a scheduler needs to differentiate packets based on their performance requirements. Either a dynamic sorted priority queue or a static priority queue can be used for this purpose. In the case when the server consists of a static priority scheduler, additional mechanisms are needed to ensure that packets with high priority do not starve packets with low priority. To avoid this problem, stop-and-go schedulers adopt a non-work-conserving multiple framing approach. However, this has the complexity of swapping the two FIFO queues at the start of each frame time.

To provide guaranteed service, the end-to-end delay bound needs to be provided in a networking environment on a per connection basis. There are various techniques to calculate the end-to-end delay bounds. One solution is to analyze the worst-case delay at each switch independently, and calculate the end-to-end delay of a connection by using the sum of the local delay bounds at all switches traversed by the connection. The traffic needs to be characterized on a per connection basis at each switch inside the network. For most of the work-conserving disciplines, due to the difficulty of characterizing traffic inside the network, tight end-to-end bounds can be derived only for rate-proportional assignments. For rate-controlled disciplines, e.g., RCSP, etc., because the traffic is regulated inside the network, tight end-to-end delay bounds can be derived for general resource assignments, and this makes the estimation of buffer space in routers easy. Scheduling schemes such as jitter-EDD assign the bandwidth at the peak rate with no statistical multiplexing. This can contribute to congestion during heavy traffic load. WFQ, on the other hand, offers better bandwidth utilization under lighter load conditions, but it assigns fixed bandwidth to sessions over a period. That is, the assignment of weights to any queue within a scheduler, or across a number of schedulers, is essentially static. This leads to high end-to-end delay under congestion conditions.

Active scheduling has addressed the drawbacks of static WFQ. Static WFQ allocates fixed weights to the sessions over a period; this leads to higher end-to-end delay and delay variation when there is high congestion on the links. Active scheduling, on the other hand, allows the introduction of a procedure by which the queuing system of heavily loaded routers is dynamically altered according to user demands. It maintains delay bounds of sessions by changing the weights of sessions according to maximum, minimum, and nominal weight limits on a per node basis. It controls the end-to-end delay of sessions by spreading the bursts over the queues of routers spanning the VPNs. Whenever there is a burst in the downstream router, the weight of this bursty session is reduced in the upstream router so that other sessions in the downstream router are served within the delay bounds. This keeps the end-to-end delays of sessions within bounds and offers better QoS. The mathematical model for the active scheduling provides delay bounds for sessions according to the weight limits of the sessions.

Exercises

1. What are the main features of a packet-scheduling algorithm?
2. Is WFQ a work-conserving or non-work-conserving algorithm? Discuss.
3. What is the difference in the workings of WFQ and WF^{2}Q?
4. How does the earliest-due-date (EDD) algorithm calculate deadlines for a packet?
5. Does active scheduling perform better than WFQ?
6. How are weights calculated in active scheduling? Why are nominal delay limits calculated in active scheduling?

References

1. S. Sezer, A. Marshall, R.F. Woods, and F. Garcia-Palacios, Buffer architectures for predictable quality of service at the ATM layer, in *IEEE Global Telecommunications Conference, GLOBECOM'98*, Sydney, November 1998, pp. 1242–1248.

2. S.A. Hussain and A. Marshall, An active scheduling policy for programmable routers, in *Sixteenth UK Teletraffic Symposium: Management of QoS — The New Challenge*, Harlow, May 2000, pp. 19/1–19/6.

3. A.K. Parekh and R.G. Gallager, A generalised processor sharing approach to flow control in integrated services networks: the single node case, *IEEE/ACM Trans. Networking*, 1, 344–357, 1993.

4. L. Zhang, Virtual clock: a new traffic control algorithm for packet switching networks, in *Proceedings of ACM SIGCOM'90*, Philadelphia, September 1990, pp. 19–29.

5. J. Golestani, A self-clocked fair queuing scheme for broad-band applications, in *Proceedings of IEEE INFOCOM'94*, Toronto, June 1994, pp. 636–646.

6. J. Bennett and H. Zhang, WF^{2}Q: Worst-case weighted fair queuing, in *Proceedings of IEEE INFOCOM'96*, San Francisco, March 1996, pp. 120–128.

7. J. Bennett and H. Zhang, Why WFQ Is Not Good Enough for Integrated Services Networks, paper presented at the Proceedings of NOSS DAZ'96, 1996.

8. D. Ferrari and D. Verma, A scheme for real-time channel establishment in wide-area networks, *IEEE J. Selected Areas Commun.*, 8, 368–379, 1990.

9. D. Verma, H. Zhang, and D. Ferrari, Guaranteeing Delay Jitter Bounds in Packet Switching Networks, paper presented at the Proceedings of Tricomm'91, Chapel Hill, NC, April 1991.

10. J. Golestani, A stop and go queuing framework for congestion management, in *Proceedings of ACM SIGCOMM'90*, Philadelphia, September 1990, pp. 8–18.

11. H. Zhang and D. Ferrari, Rate-controlled static priority queuing, in *Proceedings of IEEE INFOCOM'93*, San Francisco, April 1993, pp. 227–236.

12. A.K. Parekh and R.G Gallager, A generalised processor sharing approach to flow control in integrated services networks: the multiple node case, *IEEE/ACM Tran. Networking*, 137–150, 1994.

13. J.B. Nagle, On packet switches with infinite storage, *IEEE Trans. Commun.*, 35:4, 435–438, 1987.

14. M. Katevenis, S. Sidiropoulos, and C. Courcoubetis, Weighted round robin cell multiplexing in a general purpose ATM switch chip, *IEEE J. Selected Areas Commun.*, 9, 1265–1279, 1991.

15. J. Bennett and H. Zhang, Hierarchical packet fair queuing algorithms, *IEEE/ACM Trans. Networking*, 5:5, 675–689, 1997.

16. X. Fei and A. Marshall, Delay optimized worst case fair WFQ (WF^{2}Q) packet scheduling, in *Proceedings of ICC'02*, New York, May 2002, pp. 1080–1085.

17. P. Goyal, H.M. Vin, and H. Chen, Start-time fair queuing: a scheduling algorithm for integrated services packet switching networks, *IEEE/ACM Trans. Networking*, 5:5, 690–704, 1997.
18. H. Zhang, Service disciplines for guaranteed performance service in packet-switching networks, *Proc. IEEE*, 83, 1374–1396, 1995.
19. F.M. Chiussi, A distributed scheduling architecture for scalable packet switches, *IEEE J. Selected Areas Commun.*, 18, 2665–2682, 2000.
20. S.A. Hussain, An active scheduling paradigm for open adaptive network environments, *Int. J. Commun. Syst.*, 17, 491–506, 2004.
21. D. Stiliadis, Rate proportional servers: a design methodology for fair queuing algorithms, *IEEE/ACM Trans. Networking*, 6, 164–174, 1998.
22. S.A. Hussain and A. Marshall, A Programmable Scheduling Paradigm for Routers in an Adaptive Open Network Environment, paper presented at IASTED, Innsbruck, Austria, 2001.

Chapter 6

Active Network Management

6.1 Introduction

Traditionally, the network management level has been separated from the communication level. Different infrastructures existed for the delivery of the user data and exchange of commands between managing entities and managed objects. The situation is different these days, in particular with the Internet, where management and transport services share the communication infrastructure. In the present network management architecture, there are three main entities: a managing entity, typically a centralized application that requires human intervention; the managed objects, such as network equipment and their software; and the Simple Network Management Protocol (SNMP).[1] The work on network management standards began maturing in the late 1980s, with Open System Interconnect (OSI) CMISE/CMIP (Common Management Information Services Element/Common Management Information Protocol)[1] and the Internet- or Transmission Control Protocol/Internet Protocol (TCP/IP)-based SNMP[1] emerging as the most important standards. Both standards are designed to be independent of vender-specific products or networks. These two standards differ in their information structure, management functions, and underlying communications. The key difference between CMIP and SNMP objects is that a single CMIP object may model a complex resource. For example, CMIP objects may contain attributes to model properties of that resource, compared to SNMP object models, which represent only one property of a resource. CMIP requires a connected-oriented service, and SNMP requires a connectionless service. Today SNMP is the most widely used and deployed network management protocol, and it has been designed to operate with different vendor products and networks. SNMP is basically a centralized framework. It consists of four key elements:[2]

1. Management station or manager
2. Agent
3. Management information base (MIB)
4. Network management protocol

The manager consists of a database of network management-related information extracted from the databases of all the managed devices in the network. The software agent program embedded in a network device, such as hosts, bridges, and routers, collects traffic statistics and stores configuration values in a management information base (MIB). The manager can obtain the status information of a device or modify the configuration of the device by polling the agent. Managing agents have a defined active role; i.e., they can only send an asynchronous trap message to the manager when a few particular events occur, such as bootstrap and link failure. In SNMP framework, management agents are not able to manage faults locally; to coordinate other managing agents, they need involvement of the manager. At the Internet level, a distributed approach is followed due to a large number of nodes. This approach ensures scalability.

The current networks are closed and static systems and impose the approved definition of the entities that are to be managed; active and programmable networks, on the other hand, offer a dynamic evolution of networks. The traditional networks rely on well-defined management functionality at each management interface. With active and programmable networks, new network services and applications can be easily deployed on short notice, and hence the requirements for a management platform change. The other reason is that in active and programmable networks, users have much deeper interaction with the network because they are allowed to incorporate code intended to change behavior or configuration. This means that the network is much more exposed unless special mechanisms are adopted.

In this situation, the previous standard definition of the monitorable data has changed. It is necessary that the corresponding management layer adapt automatically, extending its initial functionality as required. The traditional SNMP and other current management protocols will not work for these new services and applications. Hence, new management frameworks must be developed to work in these environments.

The provision of fault tolerance has become a crucial feature in today's networks. Fault tolerance is commonly defined as the set of mechanisms that are deployed to cope with network failures. For a proactive network fault tolerance, it is necessary that these services are monitored by means of continuous monitoring and negative events are predicted before they occur. In this scenario, active networks can prove to be beneficial for the implementation of this requirement. In addition to the active or proactive fault tolerance, active networks provide on-the-fly data processing along the path and easy access to a broad range of information held by intermediate nodes.

Mobile agents carrying active code can be used to filter information from the MIB objects of an intermediate node in an efficient way. For example, an agent could make use of an active code to look up the MIB objects of an intermediate node and select some entries according to a need. Either it can send such retrieved information back to the application or it can use the information to make timely decisions independent from the application.[1,3] Other examples of active networks' applications in network management are congestion control, error management, and traffic monitoring. Another instance is the customization of the routing functions and scheduling schemes. A mobile agent could be used to select a best route or a scheduling scheme according to the user's quality of service (QoS) specifications. With the help of active networks, each application could set up its own control policy domain or exploit a common service (the default per hop forward function).[4] Active networks' applications can easily implement and adopt distributed and autonomous strategies in the network by sending mobile agents in the network.

6.2 Active Network Management Architectures

This section describes different active network-based architectures. Because active and programmable networks will provide complex customized services in both scale and function on a very short timescale, it will be necessary to have a highly automated and robust network management and control system. These active network management systems deal with the changing functionalities of the network nodes.[5–7] Conventional network management systems will not be able to cope with such situations. The traditional network management framework and tools are aimed at the management of static network devices and their software, which are not supposed to change frequently.

6.2.1 Application-Layer Active Networking (ALAN)

ALAN has been proposed by I.W. Marshall et al.[8,9] It is an active service architecture that provides management of the active programming of intermediate servers. It is based on users supplying Java-based active code-based services (proxylets). These services run on a selected group of servers called dynamic proxy servers (DPSs). These servers provide execution environments for proxylets (EEPs), which are executed on behalf of users. End-to-end active services are provided by one or more EEPs executing one or more proxylets. This ultimately provides an agent/management agent-based interface between network-unaware user applications and the network to make the data flows secure or QoS aware. This interface responds to high-level instructions from system administrators. Users can also utilize it to run programs and active services by adding a policy pointing to the location of their programs and offering an invocation trigger.[8,9] Messages for communication are encoded using Hypertext Markup Language (HTML) and eXtensible Markup Language (XML) and are normally carried over Hypertext Transfer Protocol (HTTP). Figure 6.1 shows a schematic diagram of a possible ALAN node.

The management system is based on role-driven policies.[9] With this architecture, users are able to specify their configuration, monitoring, and notification requirements to network devices using policies. The management policies for the user programs are incorporated in an XML metafile. Users have the liberty to separately add management policies associated with their programs using HTTP post commands. In addition, the management agent can accept policies from other agents and export policies

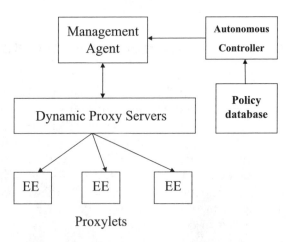

Figure 6.1 Schematic of ALAN node.

to other agents. Each policy identifies a subject (the policy interpreter), a target list (the objects to be changed if the policy is activated), an action list (the things to be done to each target), a grade of service statement, and the authorization code, ID, and reply address of the originator. The policies are named (globally unique) using a universal policy name. The management system works as an active network management system because the policies carry enclosures (e.g., the code required to execute an action, or a pointer to it) embedded in the action list. The enclosures can be instances of active services, i.e., proxylets.[8] Hence, policies are based on active service execution.

6.2.1.1 Management Agent Services

The function of a management agent on the node is to receive notifications from any entities, such as system operators and users, that send them and interpret the policy named in the notification wrapper. The management agent sends policy notifications to event service elements and notifications service elements, which are also part of a single node.

Depending on the policies supplied by the management agent, the event service elements generate events and deliver them to the notification service elements, which in turn dispatch them to notification consumers identified by the notification policies. A *notification consumer* is any entity that accepts notifications, for example, a node that implements.

The event/notification service element may receive multiple policies for the same destinations. In this case, for each receiver, the service element needs to determine which policy takes priority over the others. This priority is determined depending on the authority; for example, a policy created by a manager will usually have more priority than one created by a user.

6.2.1.2 Structures of Event Service Element and Notification Service Element

6.2.1.2.1 Event Service Elements (ESEs)

Event service is initiated when an authorized entity, such as a user or administrator, requests the monitoring and generation of events by sending an event policy to one or more ESEs.[8] The policy can contain information such as the processing details, the list of entities and attributes to be monitored, and activation, deactivation, and deletion commands.[9] Figure 6.2 shows the components of an event service element (within dotted line).

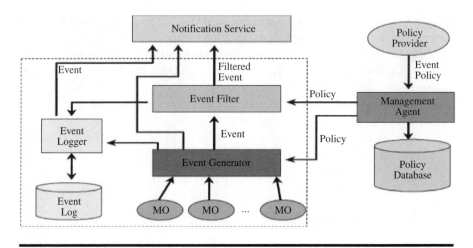

Figure 6.2   Components of an event service element. MOs = managed objects.[8]

Event Generator — The event generator polls managed objects regularly to detect changes. Alternatively, the objects may send changes asynchronously to the event generator. According to the event policy, once the event generator has formatted the change into an event, it either stores the events in the *event log*, sends them to the *event filter*, or sends them directly to the notification service. The event filter receives events and filters them according to the event policy.

Event Logger — The logger stores the events sent to it in the event database or log. The event logger maintains a time stamp, a sequence number, an event class, a time-to-live, the source object name (management agent), and the generating policy names for each event, as shown in the Table 6.1.

Policy Database — The policy database keeps active, deactivated, and deleted policies. The change of status (e.g., from active to deactivated, etc.) for each policy is stored and time-stamped.[8]

6.2.1.2.2   Notification Service Elements (NSEs)

Normally an execution policy would consist of a pointer to a proxylet rather than the proxylet itself, but the policy and the proxylet will often be part of the same notification, especially if the notification is a request to run a service originating from an end user.[8] The notification service is used for transmission of management information and associated data

Table 6.1   Information Carried by Each Event

Event Information	Description
Time stamp	Indicates event creation time
Event sequence number	Identifies each event generated
Management agent name	Identifies the agent that has generated the event
Source distinguished name	Identifies the attribute/resource about which the event is reported
Attribute type	Indicates the type of attribute about which the event is reported, for instance, integer, character, etc.
Attribute value	Value of the attribute, interpreted according to the attribute type
Time-to-live	Indicates the length of time (after event generation) that the event is valid; this is used to mark those events that lose their informational value after a period; if an event is received after its time-to-live has expired, it can be deleted without any processing
Policy class/ID	Identifies the policy resulting in the generation of the event
Requesting manager	Names the entity that has issued the policy resulting in the creation of the event; this is useful, for example, when the event recipient itself has not requested the creation of the event
Version number	Identifies which version of ESE generated the event

among management agents. This information could be event reports, policy distributions and updates, service code distributions and updates, or any combination of these. The notification service is implemented through notification service elements distributed across the system. An NSE consists of seven components (illustrated in Figure 6.3): information receiver, notification filter, notification encryptor, notification wrapper, notification dispatcher, notification logger, and notification log. Figure 6.3 shows the flow of information in a notification service element.

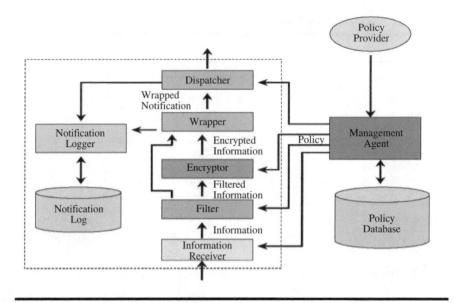

Figure 6.3 Components of a notification service element.[8]

A notification is implemented in the form of an XML entity that incorporates a policy, or a pointer to an appropriate policy if the policy was delivered before. The notifications are multicast to relevant hosts using an appropriate anycast or multicast address, where they are received by a management agent, and any enclosed policies are stored in a local policy store of the node if the appropriate key is present.[8] The management agent implements an extensible table of authorization policies to facilitate the decision of storing these policies in a local policy store.

Each notification can be considered to consist of two parts: a data part and a notification wrapper part.[8] The notification wrapper part is an XML container with the mandatory tags shown in Table 6.2.

6.2.1.3 The Autonomous Controller

The autonomous controller monitors all the execution policies in the policy database and autonomously deactivates those that are least active (as recorded in the payment log). It also informs the originator of the policy that it has been deactivated. This allows users to take necessary actions and avoid permanent deactivation if the policy has a high priority.

The autonomous controller performs two additional tasks. Policies that are never used will be deactivated completely, because nodes that possess them will be more likely to shut down.[8] Policies that are useful for some functions but not for others will exist, but may not always be active. This

Table 6.2 Notification Wrapper Tags

Notification Wrapper Tags	*Description*
Management agent name	Uniquely identifies the agent (and NSE) that has generated the notification
Notification sequence number	Uniquely identifies each notification generated by an NSE
Reply needed flag	Indicates whether the recipient is expected to return an acknowledgment
Priority (low, medium, high)	Indicates the notification's priority; this can be used to categorize the notifications and determine the resources required for their processing; for instance, high-priority notifications may be placed on high-priority output queues, and hence transmitted more quickly
Version number	Identifies which version of the notification service has generated the notification
Policy location	The policy may be part of the data enclosure or in the policy store of the receiving agent
Policy name	Points to the policy for processing the notification; any data enclosures in the notification must conform to the implicit expectations of the policy's action list; normally, the data enclosure is also an XML document

effectively means that useful active services will spread and useless ones will disappear, i.e., service deployment, configuration, and withdrawal have been automated through the autonomous controller.

The autonomous controller thus acts as a configuration manager, distributing policies/services to where they are needed and activating them on demand, without going into the details of what the demand is or what the policies represent.[8]

6.2.2 Active Networks Management Framework

A. Barone et al.[10] presented a network management framework for active networks. It manages groups of active nodes by redeploying network resources and changing nodes connectivity. In this architecture, a graphical user interface (GUI) (remote front end) of the managing entity allows the operator to design new network topologies, to check the status of the nodes, and to configure them. Additionally, the framework permits monitoring of the active applications, querying of each node, and installing of programmable traps (enabling an agent to send unsolicited notifications to the management station when some significant event occurs). Monitoring of the network devices is performed through active MIBs and agents, which are dynamic and programmable. In this framework, the managing entities communicate with an active node (the active network access point) through XML commands. An active network gateway in the active network access point performs the translation of XML responses into an active packet language and injects the code into the network. This framework also performs network failure detection by tracing network events in time.

The authors in the proposed framework introduce programmability in the management information base (MIB) and in the local agents to make the management applications active. Active applications' management consists of deployment, integration, and coordination of the software components or agents to monitor, test, poll, configure, analyze, evaluate, and control network resources. In the framework, the managing entity (ME) sends queries and receives replies in the eXtensible Markup Language (XML) to and from an active node (AN) access point, as shown in Figure 6.4. An AN access point is an active node that performs a gateway service. A gateway service performs two tasks: translation of XML requests to the specific language adopted by the execution environment (EE) and injection of the appropriate active capsules in the network that send ME requests. This way, the gateway performs the role of an interface for a particular active network, and it is specific for the language supported by its EE. Several nodes in the active network can be configured to provide gateway services.

An agent called the active local agent (ALA) is installed in each active node. The ME can poll the active local agents for get/set commands and filtering. Local agents asynchronously complete subtasks in terms of actions to be executed at local event occurrences.

The ME and gateway communicate to discover the network topology, explore the network nodes, find out which active applications are running, and monitor their activities. Additionally, the ME may include EE-specific code in XML requests to implement customized management services. A specific tag wraps user code in XML requests. User code causes safety and security problems that can be managed by means of an authentication and authorization policy. All these functions require an EE-specific code fragment stored at the gateway. This way, the gateway node provides transparent access to different active networks.

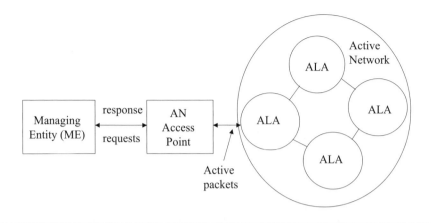

Figure 6.4   Active network management framework.

6.2.2.1   Active MIB and Active Local Agent

In the traditional SNMP framework, the local agents have only predefined capabilities and the management station can only perform basic operations of Set and Get on the MIBs, whereas the active local agent is programmable. SNMP has only limited capabilities to manage a network node or equipment that, in general, may contain several monitorable objects, either hardware or software (for instance, network interface cards and routing protocols). In contrast, in this model the ME can program the local agent's behavior to accomplish independent tasks. The MIBs are called active MIB (AMIB). These AMIBs are programmable, that is, they allow applications to store data and eventually a related code. This code is represented as a set of conditions called filters. This architecture supports distributed applications whose software components run in both end nodes and intermediate nodes of the network. An application can register with a local agent and store information in the local active MIB (AMIB). Applications can set actions to be executed by the ALA by sending incoming active capsules that can access the data and code in the AMIB, and in the action list through the services provided by the agent interface and policy enforcement module (set/get data, filters, actions). There are a variety of tasks that are performed through active capsules. For example, as a result of a queue overflow event, a warning message is sent back to the ME through a capsule, or a proactive action could be taken through a complex code in the capsule. AMIB filters can be injected by either the application itself (the data owner) or the managing entity. The network security in this architecture is of primary concern to avoid the malicious or malfunctioning codes. AMIB filters are not allowed to access network primitives. This restriction on filters and actions is implemented by a proper security policy based on authentication and authorization. ALA is supposed to enforce this security policy in the management of AMIB data, filters, and actions. The management framework can be categorized as active because (1) the MIBs are programmable, because they contain both data and code, and (2) each local management agent (ALA) plays an active role in the network management by independently performing programmed tasks, for example, policy enforcement.

6.2.2.2 Application and Service Implementation

There are a variety of management services that are implemented through this framework, for example, variable monitoring of nodes. In this service, the application can decide which data in the active MIB can be monitored by other applications, e.g., the ME. Variable monitoring can be used during the debugging activity. The other service is interapplications communication; the AMIBs allow the realization of a communication system between network applications when AMIB objects are used as shared memory variables. Interapplications communication makes new network services possible: applications can share data collected from other applications. The other example of a network service implemented is DeliveryMST. This service constructs a minimum spanning tree along which the results are progressively merged and finally delivered to the source. This service is optimal in the bandwidth consumption, and it avoids message implosion at the source. Other services that were implemented include topology discovery, monitoring the path of active packets, monitoring the routing tables, etc. Applications outside of the network domain can use the basic gateway services to obtain information on the active services available at a node and to find fault situations. Table 6.3 to Table 6.6 show the summary of services implemented in this framework.

6.2.2.3 Network Events Mining

Network events mining (NEM) deals with large archives of events obtained from the network systems and extracts useful information from these data. It also diagnoses root causes of network faults and performance degradations by establishing relationships between network events.[10]

This management framework exploits network programmability to enable management of both the active network execution environments and the active applications running on them. It enhances functionality of the classical MIB objects by converting them into more powerful entities by associating user-customizable code with the network data.

Table 6.3   Basic Gateway Services (Active Network Level: Public Services)

Type of Service	Description
GetTopology	Retrieves the topology of the current active network and information on the status of active nodes
Ping	Uses the UNIX ping command to test if a node is reachable at the network layer
ActivePing	Active version of the ping command that uses active messages to test if a node is reachable at the active network layer, i.e., at the execution environment layer
GetRoutingTable	Provides the routing table of an active node

6.2.3 FAIN: Policy-Based Network Management (PBNM) Architecture

C. Tsarouchis et al. presented the FAIN (Future Active IP Networks) project.[11] The authors developed an active network (AN) architecture geared for dynamic service deployment in heterogeneous networks. This architecture deals with the design and implementation of active nodes that support different types of execution environments, policy-based-driven network management, and a platform-independent approach to service specification and deployment.[11]

The FAIN PBNM architecture is hierarchical and two tiered. The two levels are the network management level, which encompasses the network management system (NMS), and the element management level, which encompasses the element management system (EMS).[11] There are different types of policies categorized according to the types of management operations, for example, QoS operations and service-specific operations. Hence, policies that belong to a particular category are processed by dedicated policy decision points (PDPs) and policy enforcement points (PEPs), as shown in Figure 6.5.

The NMS is the recipient of policies that may have been the result of network operator management decisions or service level agreements (SLAs) between the service provider and users. The implementation of these SLAs requires reconfiguration of the network, which is facilitated by means of policies sent to the NMS. When enforced, the policies are delivered to the NMS PEPs that map them to element level policies, which are in turn sent to the EMSs.

Table 6.4   Active Network Level: Administrative Services

Type of Service	Description
SetTopology	In an experimental test bed, defines the topology of the active network; it generates and deploys the configuration required for each active node
BootstrapAN	Spawns the EE processes in all active nodes; the service is available only if a topology configuration has been deployed by the SetTopology service
ShutdownAN	Stops the EE processes of all active nodes
SetStaticRoute	Sets a static route between a source and a destination
DeliveryPath	Generates an active capsule, which visits the active node of a given source–destination path to execute a given task; the task is a code fragment, which is provided by the user in the language supported by the EE
DeliveryPath&Return	As DeliveryPath, but in this case, the path is traversed in two directions: forwards and backwards; each node is visited twice; in the backwards path, results will be progressively merged and finally delivered to the source of the request
DeliveryMST	Generates an active capsule, which optimally visits the active network to execute a given task in each active node; the service builds a temporary minimum spanning tree along which the results will be progressively merged and finally delivered to the source of the request
DeliveryTree	As DeliveryMST, but in this case, the nodes to be visited are a subset of the network nodes

6.2.3.1   Policy Editor

The policy editor offers a GUI and a toolset in the form of templates and wizards for the composition of policies at the network level. These are generic enough to accommodate different types of policies, thus exploiting the architecture's extension capabilities.

Table 6.5   Active Application Level: Public Services

Type of Service	Description
GetApplicationsList	Retrieves the list of the active applications that have stored data in the AMIB of a given active node
GetDataList	Retrieves the list of the AMIB object IDs of a given application in an active node
GetValue	Retrieves the current value of a particular AMIB data of a given application in an active node
GetFiltersList	Retrieves the active filters associated with a particular AMIB data of a given application in an active node
GetEventsList	Retrieves the events of a given application in an active node; application events are defined in filters and actions
GetActionsList	Retrieves the actions list of a given application in an active node
GetActionCode	Retrieves the action code fragment of a given application in an active node

Table 6.6   Active Application Level: Administrative Services

Type of Service	Description
SetValue	Sets the value of a particular AMIB object of a given application in an active node
SetFilters	Modifies the active filters associated with a particular AMIB data of a given application in an active node; setting an empty list of filters is used to remove filters
Set/RemoveAction	Modifies the active actions of a given application in an active node

6.2.3.2   Active Network Service Provider (ANSP) Proxy

Policies originating at the policy editor are sent to the ANSP proxy. This has been initiated to enhance the security of the system. It provides authentication and authorization of the incoming requests (policies) and finds the management instances (MIs) to which the policies must be forwarded.

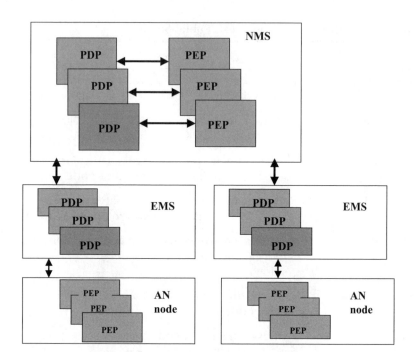

Figure 6.5   The hierarchical FAIN management architecture.

6.2.3.3   Inter-PDP Conflict Check

Conflicts may appear when a policy arriving at one PDP clashes with a contradicting policy processed by another PDP. The inter-PDP conflict check resolves complex policy conflicts by capturing inter-PDP semantics in a hierarchical manner.

6.2.3.4   PDP Manager

The PDP manager collects policies and sends them to the appropriate PDPs. If the corresponding PDP is not available, the PDP manager requests it from the Active Server Pages (ASP) framework developed by the FAIN project and installs it; in this way, it extends the management functionality of the system when needed. Different types of PDPs are incorporated into this architecture, and each corresponds to a specific context: QoS PDP, delegation of access rights PDP, and service-specific PDPs. They all perform conflict checks that are meaningful within their decision context (intra-PDP). To reach a decision, they also interact with other components that assist the PDPs in making a decision (e.g., a resource manager for admission control). The QoS PDP is responsible for analyzing QoS policies. Specifically, it decides when a policy should be deployed. It forwards decisions to PEP components to be enforced and accepts requests that come from PEPs. It also uninstalls expired policies.

6.2.3.5 Policy Enforcement Points (PEPs)

Each type of PDP has its own corresponding PEP. Network-level policies are translated by the network-level PEP into element-level policies, and then sent to the corresponding element PDPs that exist in the EMSs.

Similarly, element-level PEPs enforce the element-level policies sent by the EMS PDPs by mapping them onto the correct FAIN node open interfaces. The use of open interfaces permits all PEPs across the network to share the same view of nodes' control interfaces, making them node (platform) independent.[11]

The policies in this architecture are used as a way of managing active networks, and used as mechanisms to extend the management architecture by dynamically deploying additional PDPs and PEPs.

Although PDPs and PEPs can be deployed on demand, they must comply with the expected (standardized) interface and be registered in the ASP system. This management architecture supports the process of dynamically adding new functionality (policy action and condition interpreters) into already existing PDPs/PEPs. The concept of network management is extended by allowing multiple management architectures to be instantiated and function independently of each other.

6.2.4 Active Distributed Management (ADM) for IP Networks

R. Kawamura and R. Stadler[12] proposed active distributed management architecture, which makes use of the active network and mobile agent technologies, and thus introduces the properties of distributed control and programmability inside the network. The ADM architecture is implemented as a management middleware composed of several layers. These layers facilitate introduction of distributed algorithms that are typical for management applications and a collection of building blocks for constructing management programs.

6.2.4.1 Architecture of ADM

Figure 6.6 shows a distributed view of the ADM architecture. Management programs can be introduced from a management station attached to any of the active nodes. Active code is either installed on the nodes or

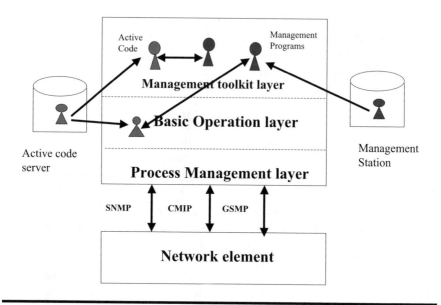

Figure 6.6   A distributed view of the ADM architecture.

downloaded on demand from an active code server.[12] The ADM middleware is divided into three sublayers, performing functions at different levels of abstraction. Following are the details of each layer's functions.

6.2.4.1.1 Process Management Layer

Process management layer implements management of active processes and provides the active networks functionality to the higher layers. This layer is present at every node and contains the most basic code of the ADM middleware. Most functions of this layer are supported by mobile agent platforms.

6.2.4.1.2 Basic Operation Layer

A management program is executed in the form of one or more active processes, which can move from node to node to execute programs. Therefore, management programs are dependent on the navigation patterns and number of active processes. (The number of active processes indicates the degree of parallelism for program execution.)

The basic operation layer provides a set of functions for navigation and different levels of parallelism for management programs. Each of these functions is implemented as an active code fragment. These navigation functions describe the degree of parallelism and internal synchronization of a management program based on the active processes associated with a specific program. The same navigation pattern can be used again for different management tasks because it is not dependent on specific management tasks. The basic operations of this layer are node control and monitoring, flow/path control, e.g., trace route, bottleneck detection, VPN signaling, and distribution of agents to all nodes in subnets for congestion location detection. These functions facilitate development of efficient and robust management programs to a great extent. This flexibility reduces the burden of developing distributed programs from the start; instead, active code fragments can achieve this functionality on very short timescale. These functions can be stored either locally or on an external active code server; from there they can be loaded on demand.

6.2.4.1.3 Management Toolkit Layer

The implementation of various network management functions is performed by the management toolkit layer using the navigation patterns provided by basic operations layer. It provides the highest level of abstraction to the application developers and supports them in writing management programs.[12] This toolkit provides high-level language constructs for network management. Like the code of layer 2, the management toolkit layer can be stored locally or downloaded on demand from an active code server.

The network management architecture described above is distributed and programmable. This management architecture is well suited for future Internet management requirements because the Internet service model is based on a simple best-effort datagram service. This represents limited environment; the network management functionality is basically restricted to element management and some diagnostic end-to-end capabilities. However, this scenario is rapidly changing. The future Internet is visualized as a multiservice platform, supporting real-time traffic and complex multiuser services with quality of service (QoS) support. It will also include differentiation mechanisms (e.g., Integrated and Differentiated Services (IntServ, DiffServ)), resource reservation schemes (e.g., Resource Reservation Protocol (RSVP)), reliability features, a security framework, virtual private networking services, and a service level agreement (SLA) architecture implementation through agents enabling interactions between Internet service providers (ISPs) and users.

The traditional network management solutions cannot meet these requirements. A property of these solutions is that they depend on a centralized hierarchical scheme for control, where management programs control and coordinate the execution of commands at the network element level, often through several levels of a hierarchy. This scheme will hardly be scalable and will not allow efficient operation for extended network environments where there is large number of nodes. An inherently distributed management architecture can solve this problem. The distributed management architecture will also provide reliability to the management infrastructure by avoiding a single point of failure.

The other characteristic of this architecture is programmability, in addition to allowing for distributed operation and control. New services and features are expected to be added on a regular basis as the Internet develops. This requires the facility of the continuous addition and extension of management functionality, which can only be facilitated by a flexible, programmable management platform. The network improvements, such as end-to-end QoS mechanisms and security functions, will require a host of new management applications, which also require a programmable management platform.[12]

6.2.5 Managing Active Networks Based on Policies (MANBoP)

MANBoP[13] is an enhancement to FAIN.[11] MANBoP has introduced decision-making logic that is independent of the network technology and the management functional domain. This reduces the logic that needs to be dynamically downloaded at runtime.[3] It facilitates interoperability and performs seamlessly operations with all types of active, programmable, and passive networks. The management architecture is also scalable.

A MANBoP-based management system consists of a single or several interconnected MANBoP modules. The MANBoP performs network configuration, active service deployment, and end-to-end network services assurance.[3]

6.2.5.1 Management System Setup

The modules constituting the system can be completely installed in management platforms or partially installed in management platforms and active nodes. This is shown in Figure 6.7, representing some modules shared between the management system and network infrastructure layers. Also shown in Figure 6.7 is an arbitrary layout linking together the modules and hence creating a particular management system. This management layout is decided by the network operator or the owner of the infrastructure and is set at the system bootstrap. Each management module knows the path to find a configuration file containing layout information, and all modules have access to configuration servers.[3]

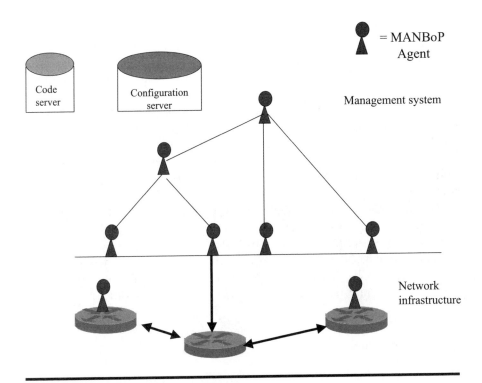

Figure 6.7   A MANBoP-based management system.

After the management system layout is decided, each module has the option to extend its functionality by downloading the appropriate code from code servers. This functional extension can also be performed at any time during the operational cycle if the module is aware that it requires a function not yet available. The network infrastructure layer in Figure 6.7 consists of active, passive, and programmable nodes. The module will be aware of the type of node it is going to manage (it receives this information reading the configuration file) and consequently will change its functionality according to the characteristics of the node by downloading the code from code servers.

6.2.5.2   The Management Module

The MANBoP module architecture consists of three categories of functional components: core, instantiable, and installable components.

The core is a set of functional components always present in the management module because they offer processing logic independent of both the functional domain and the managed technologies. Because the policy-based management paradigm[13] was embraced here, the core is actually the policy engine responsible for storage of the policies and decision making on the policy rules. Also, the core is supposed to authenticate and check user rights when users try to install their own policies, as well as the overall coordination of the MANBoP module and the downloading of code from the code server.

Installable functional components are dynamically downloaded from a code server when required by the MANBoP module. The type of installable component is selected based on three factors: the type of policy processed (i.e., its functional domain), the role played by the MANBoP management module within the system, and the underlying device technology, if any.

The instantiable functional components are need dependent. They may be needed or not; it depends on the function of the module within the management system. When the module is acting as an element-level manager directly on top of network elements, the signal demultiplexing component will be instantiated, but not if the module is a high-level management entity dealing with other management modules.[3]

6.3 Summary

This chapter provides an overview of active network management architectures. The importance of an efficient network management architecture is a reality no one can deny. Because active and programmable networks allow introduction of new services dynamically at runtime, the traditional management framework may not produce the expected results.

ALAN is an active service architecture that provides management of active programming of intermediate servers. Users supply Java-based active code services (proxylets). These services run on a selected group of servers called dynamic proxy servers (DPSs). This management system is based on role-driven policies. With this architecture, users are able to specify their configuration, monitoring, and notification requirements to network devices using policies.

A. Barone et al.[10] presented a network management architecture for active networks. It manages groups of active nodes by redeploying network resources and changing nodes' connectivity. Monitoring of the network devices is performed through active MIBs and agents, which are dynamic and programmable.

The hierarchical two-tiered FAIN PBNM architecture deals with the design and implementation of active nodes that support different types of execution environment, policy-based-driven network management, and a platform-independent approach to service specification and deployment.

The policies that belong to a particular category are processed by dedicated policy decision points (PDPs) and policy enforcement points (PEPs).

The ADM architecture is implemented as a management middleware consisting of several layers. Management programs can be introduced from a management station attached to any of the active nodes. Active code is either installed on the nodes or downloaded on demand from an active code server.

MANBoP is a further enhancement to FAIN. MANBoP introduced decision-making logic that is independent of the network technology and management functional domain. It facilitates interoperability and performs seamlessly operations with all types of active, programmable, and passive networks. The management architecture is also scalable.

Exercises

1. Discuss the structures of event service element and notification service element in ALAN.
2. What are the roles of PDPs and PEPs in FAIN management architecture?
3. Discuss the functionality of the management toolkit layer in ADM.
4. Differentiate between MANBoP and FAIN.

References

1. W. Stallings, *Computer Networking with Internet Protocols and Technology*, Prentice Hall, Englewood Cliffs, NJ, 2004.
2. E.P. Duarte, Jr., and A.L. dos Santos, Semi-active replication of SNMP objects in agent groups applied for fault management, in *IEEE*, 2001, pp. 565–578.
3. J.V. Millor and J.S. Fernández, A network management approach enabling active and programmable Internets, *IEEE Network*, 19:1, 18–24, 2005.
4. F. Safaei, I. Ouveysi, M. Zukerman, and R. Pattie, Carrier-scale programmable networks: wholesaler platform and resource optimization, *IEEE J. Selected Areas Commun.*, 19:3, 566–573, 2001.
5. M. Brunner and R. Stadler, The impact of active networking technology on service management in a telecom environment, in *Proceedings of the Sixth IFIP IEEE International Symposium*, pp. 385–400.
6. M. Brunner, B. Plattner, and R. Stadler, Service creation and management in active telecom networks, *Commun. ACM*, 44:4, 55–61, 2001.
7. B. Schwartz, A.W. Jackson, W.T. Strayer, W. Zhou, R.D. Rockwell, and C. Partridge, Smart packets: applying active networks to network management, *ACM Trans. Comput. Syst.*, 18:1, 67–88, 2000.

8. I.W. Marshall, H. Gharib, J. Hardwicke, and C. Roadknight, A Novel Architecture for Active Service Management, paper presented at 7th IFIP/IEEE International Symposium on Integrated Network Management, 2001, Seattle, WA.

9. O. Prnjat, I. Liabotis, T. Olukemi, L. Sacks, M. Fisher, P. McKee, K. Carlberg, and G. Martinez, Policy-Based Management for ALAN-Enabled Networks, paper presented at the Proceedings of the Third International Workshop on Policies for Distributed Systems and Networks (POLICY'02), 2002, pp. 181–192.

10. Barone, P. Chirco, G. Di Fatta, and G. Lo Re, A Management Architecture for Active Networks, paper presented at the Proceedings of the Fourth Annual International Workshop on Active Middleware Services (AMS'02), 2002, pp. 41–43.

11. C. Tsarouchis, S. Denazis, C. Kitahara, J. Vivero, E. Salamanca, E. Magaña, A. Galis, J.L. Mañas, Y. Carlinet, B. Mathieu, and O. Koufopavlou, A policy-based management architecture for active and programmable networks, *IEEE Network*, 17:3, 22–28, 2003.

12. R. Kawamura and R. Stadler, Active distributed management for IP networks, *IEEE Commun. Mag.*, 38:4, 114–120, 2000.

Chapter 7

Active and Programmable Routing

7.1 Introduction

Routing is the procedure of finding a path from a sender to every receiver or destination in the network. Routing algorithms calculate these paths. There are several routing protocols, ranging from simple telephone network routing protocols to complex Internet routing protocols. The terms *routing* and *forwarding* are usually used interchangeably in the networks-related literature, but these two terms are quite different in their operations. Routers deploy both of these processes. Routing is responsible for determining the end-to-end paths that packets will take from sender to destination. Routing is implemented by routing algorithms, which are responsible for filling in and updating the routing tables. A routing table contains at least two columns: the first is the address of a destination network, and the second the address of the network node that is the next hop to this destination. When a packet arrives at a router, the router controller looks up the routing table to decide the next hop for the packet. Forwarding is concerned with the localized action of a router for transferring a datagram from an input link interface to the appropriate output link interface.

Generally, a routing protocol should use minimum table space, i.e., the routing table should be as small as possible. The larger the routing table, the more will be the overhead in exchanging it. It should incur minimum overhead of control messages. A routing protocol should be robust. It should not misdirect packets.

In active routing, the instructions are carried in a capsule to select a path through the network. The conditions for selecting a path can be carried in every packet so that every packet is routed independently. There is also a possibility of carrying them in the first packet of a sequence of packets, so that a virtual circuit is set up for the duration of a connection. Active routing enables every packet or sequence of packets to use its own rules. This creates the opportunity for the coexistence of datagrams and virtual circuits on the same network, along with new routing algorithms. With this functionality network operators will select from the two options of best effort (datagram) and reserved path (virtual circuit) as the number of users who use any of these routing mechanisms change. Furthermore, new routing mechanisms can be tried without waiting for new standards.

As discussed in Chapter 2, users will have the liberty of selecting different networks and service providers through active routers that will provide end-to-end guarantees. A DiffServ mechanism is implemented in active routers to aggregate packets that require the same quality of service (QoS) or to route around network failures.

The Border Gateway Protocol (BGP) routes packets between autonomous domains, but it cannot guarantee to provide QoS or enforce security policies across networks. In active routing, the capsules carry the codes that select between the available paths to a destination. The source network can keep control of a path while the capsule is in a different autonomous domain, without having a standard method for specifying packet requirements and without having interdomain agreements. As far as the interoperability issues regarding routing in active networks are concerned, they are addressed as well. When a capsule crosses a domain, it is encapsulated in a new capsule that has code that can be interpreted in that domain. The main advantage of this type of encapsulation over the other types is that the active networks encapsulation need not to be the same for every capsule that crosses the border. The capsule can perform translation through the embedded code.

De-allocation of resources becomes more important in active networks than in the other networks. Individual users in an active network can acquire more resources of more types than in conventional networks. More importantly, the user-supplied protocols in an active network are more likely to have errors that fail to de-allocate resources than standard protocols.[1] The solution lies in the dynamic decisions by users regarding the amount of time the resources remain allocated. This amount of time is determined by the user and, in effect, becomes the soft-state refresh interval for that user. The interval may be very different for different users and may approach hard states for some users. Hard states are set up by a protocol and continue to remain until the protocol changes them. Soft states are also set by a protocol, but are automatically removed after a specified amount of time.[1] They are to be refreshed from time to time.[1]

7.1.1 Extended Label Switching

Active routing enhances the functionality of Multiprotocol Label Switching (MPLS)-capable routers. MPLS uses label-switched routers (LSRs). These LSRs are responsible for switching and routing packets on the basis of a label appended to each packet. A label is a short fixed-length identifier that is used to identify a flow of packets. MPLS uses label stacking. A labeled packet may contain a number of labels organized as a last in, first out stack.

The labels are shorter than Internet Protocol (IP) addresses because they only specify a connection in a single router, rather than a unique network destination. When a labeled packet is switched, the label is either exchanged for a label that has meaning in the next router or removed to expose the original IP packet. The exposed IP packet then uses conventional IP routing. The label-switched path (LSP) extends from the router that encapsulates the IP packet to the router that removes the encapsulation. This is like a virtual path in the Asynchronous Transfer Mode (ATM).[1]

MPLS only allows simple operations in the form of push and pop operations on label stacks that implement a single path for the portion of the route through a network-establishing tunnel. The procedures for forming stacks of labels are embedded into the router. It is hard to create new functions of labels through independent application providers. With the help of active routing, capsules can perform more difficult operations on the stacks of labels, and these capsules carry their own rules for creating new stacks. A useful extension would be conditional paths. Active routing can enable dynamic path selection instead of a single stack of labels offering a single path. Path selection can be used to select a low- or high-priority tunnel. For example, path selection can be used in a packet voice network where delay is a concern. Instead of having a single stack of labels, a capsule may carry a stack that splits at a midpoint to provide two paths to the final destination.

Active routing can extend label switching to:[1]

- Facilitate more flexible access to LSPs
- Carry out a wider range of operations on stacks of labels
- Employ new rules for constructing stacks of labels
- Allow users to establish their own label-switched paths

- Make use of more flexible restoration techniques
- Solve the problem of addressing space in IPv4

In an active network, a capsule can be used to set up a new label-switched path, as opposed to present-day MPLS, where label paths are established by the owner of the network. The remedial action for failed links can also be customized through active routing in MPLS. With active routing, each capsule may react differently at each type of failure. Some capsules may change their path, and some may send a warning message to the source.

7.2 Active Multicasting Mechanisms

7.2.1 Multicast Routing

In distributed applications, widely separated processes work together in groups. This condition generally requires that a process should send a message to all other members of the group. This type of routing is called *multicasting routing*. Depending on the size of the group, the process can send each member a point-to-point message if the group size is small. If the group size is large, but small compared to the complete network, then multicasting is used. This section has discussed the issues of reliability and fault tolerance related to multicasting.

The issue of reliable multicast, especially over the Internet, is a difficult problem. Basically, reliability means that a message is successfully delivered to all members of the group. Reliability can best be ensured if the group size is small and there are point-to-point communications among processes, but this strategy will waste network bandwidth. The concern for effective bandwidth utilization becomes more prominent when the sender and the network have a limited capacity for responding to reports of data loss or error problems. Simultaneous retransmission requests from a large number of destinations may swamp the senders and network, leading to the famous NACK (negative acknowledgments) implosion problem. The other problem pertains to the identification of packets of a selected number of receivers. Having the sender retransmit to the entire group will waste network bandwidth and bring performance down.

Another concern for reliable Internet-based multicast is the frequent group membership changes when a new process joins or leaves the group or due to process failures during communications. Such variations make it tough to specify a subset of receivers as proxies for retransmission of the lost packets. A reliable multicast is needed to deal with this scenario.

7.2.2 Active Reliable Multicast (ARM)

Lehman et al.[2] have presented a scheme called active reliable multicast (ARM) to deal with these problems. ARM uses three strategies to recover from errors. First, Lehman et al. controlled the duplicate NACKs to avoid the implosion problem. The ARM routers are deployed to reduce the amount of NACKs traffic crossing the bandwidth-starved links. Second, a recovery scheme is used to reduce the latency and to distribute the retransmissions load. The ARM routers perform caching of multicast data for possible retransmission in the upstream direction toward the source of the data packet, provided the retransmitted information is in the cache of the router. This approach facilitates quick recovery from packet losses near the receiver without imposing an unnecessary recovery overhead on the source or entire group. Third, to save bandwidth, routers are allowed to perform limited multicasting to limit the scale of retransmitted data.

Other approaches require multicast members to know the group topology or relative positions of other receivers within the multicast tree; ARM is not affected by the group topology changes because it is not dependent on any particular node (router or receiver) to perform loss recovery. The senders are ultimately responsible for retransmission of data, and the receivers are not needed to know about the group topology, and they are not supposed to buffer data for retransmission. The ARM routers perform this function.

In ARM, each data packet is assigned a unique sequence number. The packet losses are detected by sequence gaps in the data packets. This means that the receivers near the source will detect losses more quickly than the receivers farther away from the source. The receivers can also detect losses if no data arrives after a certain period.

In ARM, multiple NACKs from different receivers are cached and merged at the ARM-capable routers along the multicast tree. If all the routers are ARM capable and do not flush NACKs from their caches in a timely manner, the sender may receive at most a single NACK per packet loss. This is due to the fusion of NACKs; otherwise, the sender will receive more than one NACK. The sender responds to the first NACK by multicasting the packet again. Later, NACKs are ignored for this packet by the source for a fixed amount of time. The time is the estimated round-trip time (RTT) to the farthest receiver in the group.

There is a possibility that the NACKs and retransmitted packet are also lost; in that case, the receiver must resend a NACK if it does not receive the repair within a certain time limit. This time limit is assumed to be at least one RTT between the receiver itself and the original source of the data packet. To identify that the incoming NACK is the new NACK, each NACK contains a count to indicate how many times a receiver has requested lost data. The sender keeps the highest NACK associated with each request. If it receives a NACK with a higher NACK count, it assumes that the previous retransmission was lost and multicasts the lost packet again. Intermediate ARM routers perform the following actions by three types of packets: (1) multicast data packets, (2) NACK packets, and (3) retransmitted packets.

7.2.2.1 Data Caching for Local Retransmission

ARM routers at strategic locations cache multicast data for possible retransmission. When a router receives a NACK, indicating that a receiver has detected a packet loss, it retransmits the requested packet if it is present in its cache; otherwise, it forwards the NACK toward the sender.

7.2.2.2 NACK Fusion

Routers avoid implosion by dropping copies of the NACKs and forwarding only one NACK upstream toward the source.

7.2.2.3 Partial Multicasting for Scaled Retransmission

Routers perform partial multicasting of retransmitted packets so that they are sent to the receivers that previously requested them.

Figure 7.1 shows the header format of ARM multicast data. It has the following fields in its header:[2]

1. Multicast group address
2. Source address
3. Sequence number
4. Cache time-to-live (TTL)
5. NACK count

The cache TTL field specifies the lifetime of a data packet in an active router's cache. The NACK indicates the number of times a single receiver

Multicast Group Address	Source Address	Sequence Number	Cache TTL	NACK Count

Figure 7.1   The header format of ARM multicast data.

has requested lost packets. The amount of time a fresh data packet should be cached at a router is a function of the interpacket sending rate and the max RTT (round-trip time) between the source and the farthest receiver downstream.[2]

Additionally, an active ARM router also caches a NACK record, a REPAIR record, and likely the data packet itself. All of these items are uniquely identified by the group address, source address, and sequence number of the lost data packet.[2] A REPAIR record for a data packet indicates the outgoing links on which the repair for the data packet has already been forwarded during the time the REPAIR record was cached. The amount of time a router could cache a REPAIR record is approximately one RTT from the router to the farthest receiver downstream.

A NACK for a data packet consists of the highest NACK count received for the packet and a subscription bitmap indicating the outgoing links on which NACKs for the packet have arrived.[2] Routers use the NACK record to curb subsequent duplicate NACKs and the subscription bitmap to decide the outgoing links on which it should forward the lost or repair packet.

The ARM NACK packet has the following five fields in its header:[2]

1. Address of receiver originating the NACK
2. Address of original source of the lost data
3. Multicast group address
4. Sequence number of the requested data packet
5. NACK count
6. Cache TTL for NACK
7. REPAIR records

The NACK packet not only curbs duplicate NACKs, but also scales retransmissions. It performs scaling by checking if the request for repair or loss has been sent on that link on which the NACK arrived. If so, the NACK is dropped by the ARM router. If not, then the router looks ups its cache and if a required match is found, the router retransmits the packet. Otherwise, it forwards only the first NACK to the sender.

7.2.3 Gathercast with Active Networks

Small packets make up a large fraction of packets in the Internet. Recent studies show that 60 percent of the packets in the Internet are less than 44 bytes.[3] One example of small packets is the ACKs generated by the clients of Web servers, which are just 40 bytes long. Another example is of the Internet Cache Protocol (ICP) queries generated among Web proxies, which are about 60 bytes in length.[3]

Generally, Gathercast[3] is a network layer service that reduces the processing overhead at routers by combining small packets going to the same destination into a single packet. When combined packets arrive at the destination, they are broken into the original small packets. By combining small packets, Gathercast reduces the number of packets traveling through the intermediate routers in the network. Gathercast significantly improves network utilization by saving computation at routers.

There are several decisions that are made. These include size of packets to combine, how many packets to combine, waiting time of small packets in buffers, and at what points in the network to do this gathering. There is flexibility in choosing these parameters. For example, there is a possibility of defining small packets as 100 bytes or less, and combining just two packets into a single one has given improved performance.[3] If the delay limit is long enough, more and more small packets can be combined, but this will increase the overall latency of the packets. In addition to long latencies, this could trigger retransmissions of Transmission Control Protocol (TCP) packets. So the waiting time generally should be small. The placement of the Gathercast service is dependent on the number of senders and receivers. An increase in the number of senders will send more small packets in the network, and an addition of receivers will cause fewer packets due to aggregation in the network. Thus, the Gathercast service should generally be placed in the middle of the network.

The above-mentioned parameters are fixed, but it is unlikely that fixed values will always give the best performance. This is so because traffic in the network is not uniform; routers at different parts of the network see different traffic. When network states change, these parameters should also change accordingly based on the current conditions of the network. Thus, these values should be determined dynamically on the runtime.

Active Gathercast, due to its dynamic and programmable characteristics, is a possible solution to this problem. In this scheme, service parameters of Gathercast can adapt to network changes.

7.2.3.1 Independent Aggregators and Gatherers

Gathercast service has two functional components: aggregator and gatherer. The *aggregator* combines small packets into a single packet. The *gatherer* decomposes the combined packet and restores the original small packets.[3] A tunnel exists between the aggregator and the gatherer to ensure that the combined packets are forwarded as normal packets by intermediate routers. The combined packet is called a *Gathercast packet*.[3]

7.2.3.2 Active Gathercast Model

The aggregators and gatherers are installed dynamically into the nodes on a need basis. For other packets, the network functions as a normal IP network. This property adds great flexibility to the deployment and establishment of Gathercast. The gatherers are installed whenever they receive packets containing the protocol number IPPROTOGATHERCAST in the IP packet header. This protocol number is used to differentiate Gathercast packets from normal IP packets. The gatherers are installed onto hosts because every end node is a potential receiver.

In Gathercast, there is a possibility that a packet could be combined multiple times along its way to the destination. But this will increase aggregation times, which will eventually increase the packet latency. Because TCP is sensitive to the delay, to get the minimal delay, we choose to allow packets to be combined only once.

The aggregator's installation is triggered by a special packet. It is installed if the passing packet fulfills the following two conditions: (1) the packet meets the length limit predefined in the aggregator, and (2) the packet does not contain the protocol number IPPROTOGATHERCAST.[3] These two conditions avoid the establishment of aggregators on routers that only see large packets or combined packets. To reduce the number of aggregators, a timer is implemented on each. If there is no combination happening for some predefined threshold of consecutive small packet arrivals, the aggregator is uninstalled and the timer started. But after the timer expires, the aggregator can be activated again. In this way, dynamic aggregator installation reflects the changes of the network, and only those routers that regularly see a significant number of combinable small packets keep the aggregators.[3]

Figure 7.2 shows the whole process. In the figure, the thick arrows represent traffic with big packets; the thin arrows, traffic with small packets; and the medium arrows, traffic generated by Gathercast packets. All hosts including senders and receivers (S1 to S3 and R1 to R3) could install a gatherer. Routers A1, B1, and C1 see big packets, and thus no aggregators are installed. Routers A2 and B2 receive small packets, so periodically the aggregators are installed on these routers. Because the threshold of the number of passing small packets cannot be reached during the timer interval, the aggregator is uninstalled most of the time. Router B2 is in the middle of the network, where there are more packets. It frequently sees combinable small packets, and it keeps the aggregator deployed for a long period. Other network conditions can also be monitored by checking the IP headers of the small packets. The important information includes the average size of small packets passed during a time interval and the number of combined packets from different sources.

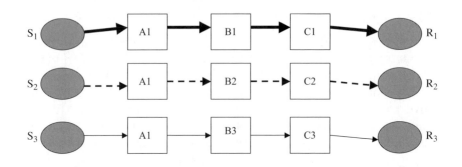

Figure 7.2 **The active Gathercast process.**

7.2.4 Hierarchical Source-Based Multicast Sessions

In traditional many-to-many networks, each sender is connected to each receiver (be it over unicast or multicast). As a consequence, a receiver has to deal with many connections and becomes overwhelmed by the amount of traffic. This does not scale well, neither in terms of bandwidth utilization needs nor in terms of computational requirements. In wireless networks, receivers such as handheld devices with little computational resources and low bandwidth on the links are severely limited in the number of connections they can handle. Thus, there is a need to assign priority to applications in many-to-many communications when the priorities of the senders are not of equal importance to the receiver. The aggregated hierarchical overlay network concept presented by Wolf and Choi[4] dynamically aggregates the information sent by certain sources, while for others it keeps full stream. This scheme provides different levels of aggregation to the end users. It also allows a user to change to a higher or lower level of detail if necessary.

As shown in Figure 7.3, each layer in the hierarchy corresponds to a different aggregation level. The lowest layer, i.e., layer 0, contains the data sources at which terminals or users send their unaggregated data. Each node in layer 1 aggregates multiple layer 0 sources to a new stream. This stream is delivered upwards to layer 2, where it is aggregated with other layer 1 streams. This continues up to the root node.

A user who is interested in getting information from a node on a certain layer can directly connect to that node and receive its aggregated data stream. Depending on the amount of information required, the user could connect to a child of the current node if more detail is needed and to the parent if less detailed information is required. The multicasting is performed through standard schemes such as International Group Management Protocol (IGMP). Multiple users can connect to a certain node

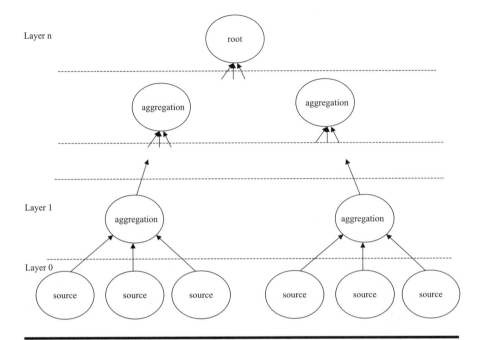

Figure 7.3   Hierarchical aggregated levels.[4]

in a layer. To ensure scalability, any node in a layer can present its data as a multicast session.

7.2.4.1   Scalability Issues

In this scheme, nodes are configured to form the tree and aggregate the right set of lower layer streams. Therefore, all nodes other than leaves need to know their children in the beginning. Once the children are known, the node can subscribe to their multicast sessions and aggregate the information.[4]

The control information that is distributed to nodes for each session includes two components. One component is the list of children and their respective session identifiers. The other is the session identifier of the parent of the current node, which is added to allow an observer to move to a higher aggregation layer.

To ensure scalability, a naming scheme has been introduced. The control data of a session cannot contain the names/identifiers of all parents, children, grandchildren, etc., because this would cause an unmanageable amount of information. Each node is identified by its name and the concatenation of names of nodes in the path from the root to the node. This results in a hierarchical naming scheme similar to that used in the

Domain Name Service. This scheme is comparable to Concast, described by K.L. Calvert et al.[5]

Although Concast implements a general-purpose many-to-one communications paradigm, it does not specifically consider timeout issues due to packet loss, and thus it is not suited for real-time applications. In the hierarchical overlay network, however, there is a mechanism to detect packet loss.

7.3   Active and Programmable Router Architectures

Routers in the future Internet scenario will not only forward data packets, but also provide value-added QoS-based services. Such a trend is motivated by both customer needs and technological advancements. The additional issues of price, security, management, and accounting will also be considered, along with the forwarding of packets in the design of routers. As the data network contents become priced and operationally advanced, network users (both service providers and users) will demand better service assurances in their transport. Security services such as copyright management, key management, and intrusion detection will be important. Accounting services allow network usage to be correctly charged. Services for quality of service (QoS) reservation, differentiation, and adaptation will allow information to be delivered with user-centric performance guarantees. Keeping in view these requirements, researchers are making efforts to design programmable routers to support these value-added services. This section of the chapter discusses different programmable router structures.

7.3.1   Flexible Intra-AS Routing Environment (FIRE)

A routing protocol has three component functions: it defines a set of metrics upon which routing decisions are made, it disseminates this information throughout the network, and it defines the algorithms that decide the paths packets use to cross the network. Additionally, a well-designed protocol contains security mechanisms to protect the routing infrastructure from attack and misuse. In today's routing protocols, these functions are closely integrated and cannot be separated. This information that is distributed about each link, and the algorithm that is used to select paths, is fixed; the operator is largely unable to change the system to use a new algorithm or different metrics.[6]

The flexible intra-AS routing environment (FIRE)[6] provides more freedom to control a variety of key routing functions, including choosing algorithms to select paths, the information to be used by the algorithms, and traffic classes to be forwarded according to the specified algorithms. Based on these concepts, FIRE has introduced programmability into the routing protocol. FIRE divides the standard routing protocol into its constituent parts: secure-state distribution, computation of forwarding tables, and generation of state information (i.e., determining what values to distribute). FIRE then depicts its state distribution functionality, making computation of forwarding tables and the generation of state information at runtime on the fly.

Most routing protocols make use of hop counts to approximate the cost of a link. More complex protocols like Extended Interior Gateway Routing Protocol (EIGRP) compute the routing metric from a mix of several link properties, but the EIGRP equation for combining metrics is fixed and represents a balance between possibly conflicting values.

FIRE supplies a set of properties for use by routing algorithms. Properties can be configured values extracted from router MIBs, or even dynamically generated by operator-provided applets. Properties can be regenerated when conditions suggest that forwarding tables need to be updated.

The increasing awareness about QoS and heterogeneity of Internet traffic strongly suggests that there is a growing diversity in path choices. Different paths between two points will be better suited for different applications. Most routing protocols in use today forward all traffic based upon the same forwarding table. Traffic classes may be differentiated with respect to resource reservation[7] and queuing priority,[8] but packets are generally routed identically. In FIRE, packets are routed based upon a forwarding table constructed to best suit its particular traffic class requirements. By assigning different classes of traffic to separate forwarding tables, FIRE permits network operators to optimize performance for each class. Each forwarding table is constructed with a different algorithm, which may utilize its own set of metrics.

7.3.1.1 Architecture and Functions

In traditional networks, each router generates a single forwarding table for itself, which the router then uses to decide where to forward incoming traffic. FIRE provides extended functionality by generating a set of forwarding tables, each uniquely characterized by three components: the algorithm selected to compute the table, the properties used by the algorithm in its computations, and a packet filter that determines which classes of traffic use the forwarding table. In FIRE, the network operator may configure all three of these variables at runtime.[6] FIRE tries to provide flexibility and security for dynamic routing infrastructure by keeping the role of a network operator more prominent. This is opposed to capsule-based active networks, which provide Java-based general execution environments to mobile code, or limited remote router configuration mechanisms through users' interaction. The three components of the programmable routers are discussed in detail below.

7.3.1.1.1 Algorithms

The routing algorithms in FIRE are downloaded Java-based programs. These algorithms generate local forwarding tables on a local router by employing distributed network properties. Each instance of an algorithm is run in a separate Java Virtual Machine (JVM) on the router itself. Java sandboxing is used to prevent a malicious routing algorithm from rendering the entire router out of order.

7.3.1.1.2 Properties

Property values are the attributes or variables that are used by routing algorithms in calculations of routes. Properties differ from metrics used in conventional routing algorithms. Metrics are weights, assigned to a link, that influence the link's chance of inclusion in forwarding paths. Properties, as opposed to metrics, are applicable not only to links, but also to routers and networks.

Some property values may be configured into the nodes (e.g., multicast support or policy-based cost values). Others may be readily available from the router's MIB (e.g., average queue length, central processing unit (CPU) utilization, etc.).[6] FIRE also permits operators to write their own property applets. Like algorithms, property applets are written in Java, and each is executed in its own JVM instance.

All property values are delivered to every node in the network using reliable means. Each node stores these values in a property database that contains property values for each entity. This builds a complete and consistent view of the network. Updates of the values of these properties and other routing information are periodically flooded throughout the network through the use of state advertisements (SAs) or state agents, refreshing the databases at each router.[6]

Certificate SAs are used to distribute public keys and authority certifications. FIRE employs a certificate infrastructure with a layered authority structure. These certificates advertise the public keys of nodes, links, and entities such as the operator. All SAs are digitally signed to provide end-to-end authentication and integrity. Additionally, FIRE makes use of IPsec to protect against hop-by-hop attacks. Routes managed by protocols other than FIRE (such as exterior routes) are advertised through the use of external-route SAs.

A cache of all current SAs is maintained by each FIRE router. The purpose of the state distribution mechanism is to maintain a consistent shared view of the set of current SAs across all routing nodes. Reliable flooding is used to make sure that an SA generated by one node is eventually received by all nodes in the SA. In addition to being time-stamped, each SA is also given a time-to-live, after which it is considered invalid and is no longer flooded by FIRE.

Upon receipt of a state message, a router first checks its cache to see if it has already been present. If so, it simply acknowledges the message and completes processing. If, on the other hand, the message contains an SA the router has not seen before, it first validates that the SA header is properly formed. If the header is invalid for whatever reason, the router immediately sends an acknowledgment to the sender, sending back the header information. Routers receiving acknowledgments compare the enclosed header with outstanding transmissions. If the SA was damaged in transmission, the headers will not match and the SA will be retransmitted as if unacknowledged.[6]

7.3.1.1.3 Filtering

Filtering or classification here refers to separation of traffic into classes. A single instance of FIRE can be used to manage several forwarding tables (a set of forwarding tables), and thus serve several classes of traffic concurrently. Network traffic is classified through operator-specified filters and forwarded according to the generated forwarding tables. This characteristic creates a set of virtual private networks (VPNs) over a single network.

Figure 7.4 shows the internal architecture of a FIRE router. In the router's data path, all incoming traffic on the router first passes through a main filter that separates the traffic by assigning it to a particular instance

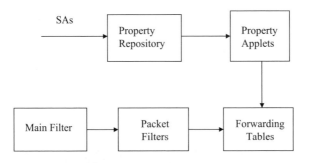

Figure 7.4   The internal architectue of a FIRE router.

of FIRE. A packet can belong to only one instance. Within the instance, different packet filters determine which forwarding table the packet will use. The packet's destination address is looked up in the indicated table, and the packet is forwarded based on that.

7.3.1.2 Configuration and Management

The operator in FIRE is allowed to configure the network through two mechanisms: the operator configuration message (OCM) and the operator configuration file (OCF).

7.3.1.3 Configuration Messages and Files

The OCM is a special SA that contains the configuration rules for the FIRE system and the set of OCFs that are to be loaded, and names one of them as the *running* OCF. An OCF records the routing algorithms to be used, the properties to be advertised, and the filters to map traffic classes to forwarding tables. To ensure integrity, both the OCMs and OCFs are signed with the operator's secret key. The OCM is sent into the network by an operator at any FIRE node and is distributed along with normal routing updates throughout the network by the standard flooding mechanism.

Upon receipt of the OCM, a node retrieves the listed OCFs from one of the file sets specified in the OCM. File retrieval is performed by a special simple File Transfer Protocol called the Large Data Transfer Protocol (LDTP).[6] This protocol is based on the Trivial File Transfer Protocol (TFTP),[9] with the improvement to protect against security threats.

7.3.2 Darwin-Based Programmable Router Architecture

Today's routers and switches implement packet classification and priority-based scheduling and congestion control mechanisms. They also offer standardized QoS mechanisms such as differentiated services, and Multiprotocol Label Switching (MPLS) users could utilize these mechanisms to handle their traffic in specific ways. The other example is value-added services, that is, services that require not only communication, but also data processing and access to storage. Because routers are closed systems with a set of standard protocols, it becomes difficult to realize these enhancements.

To realize these enhancements, J. Gao et al.[10] presented a router architecture in which the control plane functionality of the router can be extended using *delegates*. Delegates are code segments that implement customized traffic control policies or protocols. Delegates are executed as Java threads inside the virtual machine sandbox. Delegates can help the router treating the packets belonging to a specific user, such as value-added service providers, through the *router control interface* (RCI).

There are two types of operations on data flows inside the network. The first type includes manipulation of the data in the packets, such as video transcoding, compression, or encryption.[10] Because most routers are limited in general-purpose processing power, this type of processing cannot take place on them. The second type of operations is based on how the data is forwarded, but typically does not require processing or even looking at the contents of packets. Examples of this type are tunneling, rerouting, selective packet dropping, and altering the bandwidth allocation of a flow. These operations, due to their nature, are best carried out on routers or switches. Data delegates perform data processing operations and execute on servers. Control delegates execute on routers and are involved in the control of data flows.

Figure 7.5 shows a router architecture that functions through the use of delegates. The top part of the architecture is a control plane that executes control protocols such as routing and signaling, and the bottom part is a data plane that is responsible for packet forwarding. Control delegates execute in a Java runtime environment that is part of the control plane. Control delegates operate on the data through the RCI. The RCI provides a set of operations on flows of

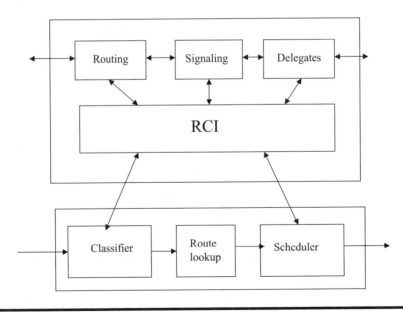

Figure 7.5 The router architecture using delegates.[10]

packets. The classifier uses the flow specifications to determine to which flow incoming packets belong.

Following are the five categories of functions that are necessary for the RCI to support a broad spectrum of delegates:

> **Flow manipulation methods**: This type of method permits delegates to define and manage flows by updating the classifier data structures. There is a programming model available to delegates. For example, a flow defined with a flow specification (a list of header fields) can be added to the classifier through the add_flow call.[10] Operations on this flow, such as specification of QoS parameters and rerouting, are then performed. Methods for deleting and modifying flow specifications are also available.
>
> **Resource management methods:** Delegates decide how resources are allocated and distributed across flows by modifying states in the scheduler through the RCI. The scheduler uses its hierarchical resource tree to schedule packets to meet each flow's QoS requirement. RCI methods of this type consist of adding nodes to and deleting nodes from the resource tree, modifying the service parameters of a node, and retrieving the resource hierarchy.[10]
>
> **Flow redirecting methods:** This type of method has a global operation strategy and changes the traffic distribution in the network, as opposed to the above class of methods, which have only local significance. For example, delegates use the reroute method to route a flow's packets through a route other than the default to avoid congestion in the network. Packet tunneling and selective dropping methods are also implemented in this category.
>
> **Traffic monitoring and queue management methods:** Delegates monitor network congestion status by monitoring queue lengths. The hierarchical scheduler implements a fairly complex queuing discipline, and RCI exports basic queue management methods to delegates.[10] For example, the method probe returns the queue size of a flow. The method retrieve_data enables delegates to retrieve bandwidth usage and delay data of each flow.
>
> **Support for delegate communication**: This method supports communication between delegates on routers in the network and also interacts with applications' end systems. This type of messaging between them allows delegates to gather global information so that proper global actions may be taken, such as rerouting for load balancing and congestion control.

7.3.3 *Programmable Router Operating System*

The operating system (OS) plays a very important role in the design of an active and programmable router. The programmable OS will make it easier for value-added services to be implemented as general programs that can be deployed on demand, but there are new challenges in the design of router OSs. First, router programs will need to access distinct system resources, such as forwarding network bandwidth, router CPU cycles, state-store capacity, etc.

Second, there will be a large number of diverse resource consumers (different applications) whose resource requirements will have to be either coordinated to enable cooperation or arbitrated to resolve competition. This requires that proper threading and scheduling techniques should be in place before any initiative for the value-added services is taken.

Third, a router OS must concurrently support multiple virtual machines exporting different APIs. This gives various benefits. For example, different APIs can be provided for different classes of applications, or different Internet service providers can prefer different router service providers. This will also enable a smooth introduction of a newer version of JVM while an older version is still working. The idea will be to not disrupt legacy services.

Yau and Chen[11] have presented an operating system called core router operating system support (CROSS).

7.3.3.1 *Operational Architecture of CROSS*

CROSS flexibly binds resource allocations at program execution time to resource consumers, namely, threads, flows, and address spaces. CPU, network, disk, and memory schedulers in CROSS multiplex[11] these allocations onto the physical router resources.

Virtual machines supported on CROSS may relate to different execution environments (EEs). CROSS supports both trusted and untrusted router programs. Trusted programs may run in the same kernel address space as CROSS, achieving tight coupling. On the other hand, untrusted programs must run in traditional OS address spaces to achieve fault isolation.

Unlike conventional OS, CROSS programs are typically invoked by asynchronous packet arrivals. Such a packet that invokes a CROSS program is traditionally called an *active* packet. A CROSS active packet carries in its IP header an option containing the call function names and parameters to be dispatched at a target router virtual machine.

The process of CROSS program dispatch is shown in Figure 7.6. When a packet arrives at an input link, a packet classifier decides if the packet is active and destined for the receiving router. If it is, we must

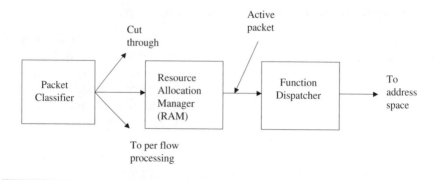

Figure 7.6   The process of CROSS program dispatch.

determine a resource allocation to use for the requested program execution. The task is achieved by a *resource allocation manager* (RAM). In the case that a resource allocation is returned by RAM for an active packet, the allocation will be passed to the CROSS *function dispatcher* (FUND) together with the call function name and parameters. FUND unwraps the call parameters. It then dispatches the call by allocating either a thread or an address space for the call context, depending on whether the function is trusted.

7.3.3.2   Packet Classification

The packet classifier is used for flow differentiation at the *application* level, packet forwarding, and resource binding. This leads to two requirements. First, some transaction-type application flows may be short-lived, causing highly dynamic updates of the lookup database. Hence, although lookup speed will remain an important factor affecting classifier performance, the efficiency of database add and delete operations will also become highly important. Second, because users can start many applications at the same time, and certain applications may even be generated automatically, the lookup database for a busy router may become quite large. Performance that scales well with the number of database entries is thus an important goal.

The main aim of resource allocations is to give CROSS programs fine-grained control regarding system resources. This enables QoS guarantees and discrimination according to application requirements and priorities. Because resource characteristics differ, scheduling algorithms must be designed in a resource-specific manner. For example, CPU context switching is expensive compared with switching between flows in network scheduling. Hence, efficiency of CPU scheduling improves if threads can receive a minimum CPU share before being preempted. Disk scheduling, unlike both CPU and network, must consider request locations to limit seek time. Memory scheduling must estimate the current working sets of resource consumer applications. Schedulers in CROSS must therefore examine relevant resource states in addition to QoS specifications.

7.3.4 *Active Routing for* Ad Hoc *Networks*

Ad hoc networks are wireless multihop networks whose continuous topology changes make the design and operation of a standard routing protocol difficult. It is impossible to find a best ad hoc routing protocol, i.e., the protocol that is suitable for all circumstances. There are many considerations, for example, link characteristics, quality of service (QoS) requirements, and security concerns, that need to be considered.

C. Tschudin et al.[12] have presented an active network-based ad hoc network routing scheme. They are of the view that routing logic can be defined and deployed at runtime to adapt to special conditions arising due to topology changes. The active placement of routing protocols facilitates users to choose the best protocol candidate depending on the network's current topology, packet arrival patterns, etc. In this case, the active network supports customization of routing protocols; rather than relying on a common routing service, users can execute their own routing protocol, actually creating a private virtual (routing) network.

7.3.4.1 *The Simple Active Packet Format (SAPF)*

This scheme uses a special header format for demultiplexing traffic from different types of forwarding mechanisms, as shown in Figure 7.7. This format is called simple active packet format (SAPF). It consists of a single *selector* field based on which router can make fast forwarding decisions. The selectors point to the respective execution environment for those active packets depending on the type of application. The SAPFs are also used to achieve interoperability between different types of

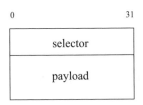

Figure 7.7 Simple active packet format.

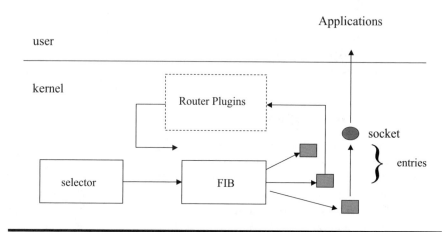

Figure 7.8 FIB's role in the working of an SAPF base.

forwarding modules by tunneling IP packets inside the SAPF. Actually, SAPF aims to reduce the forwarding module complexity to a minimum, leaving all configuration aspects to active packets.

The forwarding information base plays a key role in the working of the SAPF base, as shown in the Figure 7.8. The selector of an incoming active packet is used as an index into the forwarding information base (FIB), where different entry types can be found. FIB is a set of forwarding instructions that conditionally forward packets. Plain forwarding (FWD) entries let a packet be copied to one or more remote nodes, possibly rewriting the SAPF selector if needed. Entries in the FIB can also point to operating system-dependent data structures like sockets, from which applications can read the packet's payload. Another possibility is to let FIB entries point to special router plug-ins that process a packet before it is handed back to the SAPF forwarding engine. Lastly, FIB entries can also be branch-and-forward instructions, which conditionally forward packets along two different paths.[12]

There are entries in the FIB that are stamped with an expiry time stamp, after which the entry will be removed by the system (soft state). This is not true for a socket that is only removed when the socket is destroyed.

The basic functionality of the SAPF layer is that of forwarding datagrams. This forwarding can be redefined and redirected in arbitrary ways. This in contrast to the User Datagram Protocol (UDP), where a datagram carries an end node address; an SAPF header only selects the entry in the next downstream node. Forwarding behavior is defined on the fly by active packets that put state into the FIB.

7.3.4.2 Neighbor Discovery

The neighbor discovery is performed through active packets. The active discovery packets collect this connectivity information and store it on neighbor nodes. The neighbor discovery packets also broadcast interface names and link status information.

Each node in the ad hoc network periodically broadcasts an active *beacon* or *hello* packet whose function is to announce node's presence. This is achieved by placing the connectivity information in the persistent data storage area of all neighbor nodes. The most significant information is the source node's and outgoing interface's names. At arrival, the beacon packet will also add a time stamp (expressed in local time) to cater for the expiry of the old data. The time-stamp information also avoids clock synchronization problems among nodes.

7.3.5 Component-Based Active Network Architecture

S. Schmid et al.[13] have presented a component-based node architecture called LARA++. This architecture provides a flexible composition framework for the deployment of simple and light active software components. The presenters are of the view that the capsule-based approach for programmability of routers offers limited functionality. This is due to the size restrictions of capsules and fixed programming interface offered by the execution environments. The LARA++ take advantage of the benefits of component-based architectures, namely, code modularity, reusability, and dynamic composition, to facilitate development and deployment of customized network services.

LARA++ is a layered architecture. It consists of an active network layer on top of an existing router operating system. The architecture is *standard* in the sense that it can be implemented on any underlying router platform. Figure 7.9 shows the LARA++ layered architecture:[13]

The NodeOS offers controlled access to low-level local resources and other services such as device configurations. The main components of the NodeOS are the packet interceptor and injector, packet classifier, memory mapping system and system trap control mechanism.[13]

The packet interceptor (and injector) intercepts (and injects respectively) the data packets either on the data link layer or on the network-layer level. The packet classifier decides whether a packet needs active computation and through which active component(s).

The memory mapping system prevents expensive copy operations when passing the data into user space for active processing and back down again by mapping the physical into the virtual address space of the

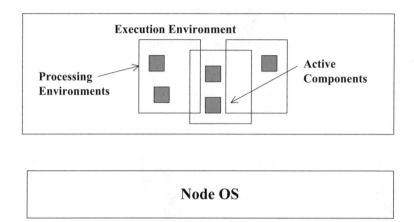

Figure 7.9   The LARA++ layered architecture.

active components involved in packet processing. The system trap control guarantees node safety by checking whether system calls to the NodeOS have been activated through the LARA++ system API.

Execution environments (EEs) enforce management and security policies on each active program within an EE. *Processing environments* (PEs) provide a safe code processing environment where active components are executed. Lastly *Active Components* consist of the real code. They are processed within PEs beside other trusted components.

7.3.5.1   Service Composition

Services in LARA++ are created through the addition of packet filters into the *packet classifier*. Active components can register their filters with the packet classifier at instantiation time or again at run-time if required.[13] The classifier decides based on the set of packet filters currently installed whether or not a packet requires active processing, which active component(s) are needed and in which sequence.[13]

The fact that active services are composed through insertion or removal of packet filters makes the component bindings dynamic. Since the service composition depends on the actual data in the packets, the bindings are conditional. The packet classifier keeps a flexible classification graph, to structure the packet filters of the active components instantiated on the router.[13] Each node in the graph comprises of a set of packet filters that define the sub-branches of the node.[13] A packet that is matched by a filter is passed to the cor-responding active component. After conclusion of the active processing, the classifier continues at the same point in the classification graph.

7.3.5.2   Processing Environments

LARA++ processing environments offer a safe execution environment for active components, through code signatures and user authentication, to protect the active node from untrusted machine code based on the principles of virtual memory management, preemptive multitasking, and system call control. They define the visibility boundaries for active code that prevent malicious code from breaking out of the protection environment. PEs also provide *efficient* active code execution environments by using a low-latency user-level thread scheduler for the PEs.[13]

7.3.5.3   Active NodeOS

The LARA++ *Active NodeOS* provides access to low-level system service routines such as device and resource configurations. It controls access for high-level components based on their privilege level and policies. The Active NodeOS forms an independent layer of service on top of the router operating system.

The main components of the LARA++ Active NodeOS are the packet interceptor/injector, classifier, memory mapping mechanism, and system trap control:[13]

■ Depending on the active network service, data packets (or frames) can be intercepted or injected either on the link-layer level or on the network-layer level.

■ The *packet classifier* decides whether a packet needs active computation and by which active component(s).

■ The *memory mapping mechanism* avoids expensive copy operations when passing the packet data into user space for active processing and back again by mapping the physical memory directly into the virtual address space of the active components that have to process the packet.

■ The *system trap control mechanism* ensures node safety by restraining the system call interface. It checks whether system calls to the Active NodeOS (or the router OS) have been invoked through the LARA++ system API.

7.4 Summary

This chapter has described the active multicasting and active router architectures presented by different researchers at different sites. It explains the concept of multicasting and then presents the schemes: active reliable multicast (ARM), Gathercast with active networks, and hierarchical source-based multicast sessions. Active reliable multicast (ARM) presents solutions for the problems of duplicate NACKs, latency due to recovery overhead, and bandwidth wastage due to retransmissions. ARM uses three strategies to recover from these problems. First, it controls duplicate NACKs to avoid the implosion problem. The ARM routers are deployed to reduce the amount of NACK traffic crossing the bandwidth-starved links. Second, a recovery scheme is used to reduce the latency and to distribute the retransmissions load. The ARM routers at specific locations perform caching of multicast data for possible retransmission in the upstream direction toward the source of the data packet. Third, to save bandwidth, routers are allowed to perform limited multicasting to limit the scale of retransmitted data. Gathercast reduces the processing overhead at routers by combining small packets into a single packet destined to the same receiver. When combined packets arrive at the receiver, they are broken into the original packets. By combining small packets, Gathercast reduces the number of packets traveling through the intermediate routers in the network. This significantly improves network utilization by saving computation at the routers.

The aggregated hierarchical overlay network concept solves the bandwidth consumption by dynamically aggregating the information sent by high-priority sources, while for others it keeps full stream. This scheme provides different levels of aggregation to the end users. It also allows a user to change to a higher or lower level of detail if necessary.

The flexible intra-AS routing environment (FIRE) allows controlling of a variety of key routing functions, including choosing algorithms to select paths, the information to be used by the algorithms, and traffic classes to be forwarded according to the specified algorithms. FIRE has introduced programmability into the routing protocol. FIRE divides the standard routing protocol into its constituent parts: secure-state distribution, computation of forwarding tables, and generation of state information (i.e., determining what values to distribute).

The Darwin-based router extends control plane functionality of the router by using code segments called delegates. This implements customized traffic control policies or protocols. The customized policies are flow manipulation methods, resource management methods, and traffic monitoring and queue management methods.

CROSS flexibly binds resource allocations, namely, threads, flows, and address spaces, at program execution time. CPU, network, disk, and memory schedulers in CROSS multiplex these allocations onto the physical router resources.

The active network-based ad hoc routing scheme allows deployment of routing logic at runtime to adapt to special conditions arising due to topology changes. The active placement of routing protocols facilitates users to choose the best protocol candidate depending on the network's conditions.

LARA++ is a component-based architecture for active network nodes. It introduces flexibility and extensibility into the active routers.

References

1. N.F. Maxemchuk and S.H. Low, Active routing, *IEEE J. Selected Areas Commun.*, 19:3, 552–565, 2001.
2. L.-W.H. Lehman, S.J. Garland, and D.L. Tennenhouse, Active reliable multicast, *IEEE*, 899–904.
3. Y. He, C.S. Raghavendra, and S. Berson, Gathercast with Active Networks, Proceedings of the Fourth Annual International Workshop on Active Middleware Services (AMS'02), 2002.
4. T. Wolf and S.Y. Choi, Aggregated hierarchical multicast for active networks, 2001, *IEEE*, 899–904.
5. K.L. Calvert, J. Griffioen, B. Mullins, A. Sehgal, and S. Wen, Concast: design and implementation of an active network service, *IEEE J. Selected Areas Commun.*, 19:3, 426–437, 2001.
6. C. Partridge, A.C. Snoeren, W.T. Strayer, B. Schwartz, M. Condell, and I. Castiñeyra, FIRE: flexible intra-AS routing environment, *IEEE J. Selected Areas Commun.*, 19:3, 410–425, 2001.
7. B. Braden, L. Zhang, S. Berson, S. Herzog, and S. Jamin, Resource Reservation Protocol (RSVP): Version 1 Functional Specification, IETF RFC 2205, September 1997.
8. K. Nichols, V. Jacobson, and L. Zhang, A Two-Bit Differentiated Services Architecture for the Internet, IETF RFC 2638, July 1999.
9. K. Sollins, The TFTP Protocol (Revision 2), IETF RFC 1350, July 1992.
10. J. Gao, P. Steenkiste, E. Takahashi, and A. Fisher, A programmable router architecture supporting control plane extensibility, *IEEE Commun. Mag.*, 38:3, 152–159, 2000.
11. D.K.Y. Yau and X. Chen, Resource management in software-programmable router operating systems, *IEEE J. Selected Areas Commun.*, 19:3, 488–500, 2001.
12. C. Tschudin, H. Lundgren, and H. Gulbrandsen, Active routing for ad hoc networks, *IEEE Commun. Mag.*, 38:4, 122–127, 2000.

13. S. Schmid, J. Finney, A.C. Scott, and W.D. Shepherd, Component-Based Active Network Architecture, paper presented at Sixth IEEE Symposium on Computers and Communications (ISCC'01), *IEEE*, 2001.

Chapter 8

Active Wireless and Mobile Networks

8.1 Introduction

Recently there has been a proliferation of ubitiqious wireless devices that offer anytime and anywhere network access. However, there is an anomaly between the characteristics of wireless mobile networks and the wired infrastructure. The main characteristics of wireless mobile networks are relatively lower bandwidth, irregular connectivity, and higher error rates. On the other hand, wired networks offer high bandwidth, steady connectivity, and very low error rates. Real-time applications that are quality of service (QoS) aware require special handling to overcome the limitations of wireless mobile networks and the interconnection of wireless and wired networks to provide up-to-the-mark performance to the user. The only solution is the provision and deployment of new and tailored services in wireless networks to meet user needs. Unfortunately, the current network infrastructure, which is fixed, cannot provide this solution, and introducing new protocols in the network requires a time-consuming standardization process.

As the research on wireless networks progresses over the next several years, there will be an increasing demand for new mobile devices, services, and radios that can meet the needs of mobile users. Recent trends indicate that a wide variety of mobile devices will be available, each requiring specialized services and protocols. The users of these services will be very conscious about quality of service (QoS). Existing mobile and wireless networks have limited service creation environments. Typically, the service creation process in these networks is manual, ad hoc, costly, and time-consuming. Mobile network services and protocols cannot be easily enhanced or modified because they are generally implemented using dedicated firmware, embedded software, or constitute part of the low-level operating system support. For example, it is difficult to dynamically modify the handoff prioritization strategy in cellular networks and wireless local area networks (WLANS) or to introduce new handoff control algorithms (e.g., mobile-assisted handoff) in WLANs. Handoff detection mechanisms are needed to find the most suitable access points (for example, base station in case of cellular networks) to which a mobile device should be attached. Wireless access points can be selected based on different factors, including channel quality measurements, resource availability, user-specific policies, and heterogeneous environments that support fundamentally different mobility management systems.[1] The incompatibility of signaling systems and physical-layer radio technologies prevents mobile devices from roaming seamlessly between heterogeneous wireless networks. For example, compatibility problems arise when a code division multiple access (CDMA) cellular phone is to be connected to an IEEE 802.11 wireless LAN. Additionally, access network protocols make specific suppositions about the capability of mobile devices and generalize these suppositions. For example, both mobile Internet Protocol (IP)[1] and wireless Asynchronous Transfer Mode (ATM)[1] assume that handoff control is located at the mobile device. Such mobile-controlled handoff schemes may not work well for many low-power devices that are incapable of continuously monitoring channel quality measurements due to their power limitations.

Active wireless networks address the above-mentioned issues by introducing adaptive services. Active nodes in these networks enable deployment of custom services and protocols. Active networking services address bandwidth limitations and physical radio layer impairments by permitting applications to install filters at wireless gateways. This chapter discusses different active and programmable wireless network approaches. For example, the approach adopted by A. Boulis et al.[2] allows application-specific customization and adaptation of packet processing at the base station and the end nodes at runtime to accommodate the diverse mobile nodes with differing capabilities. Similarly, the approach by Michael E. Kounavis et al.[3] allows programmable handoffs to cater to seamless mobility of devices across heterogeneous environments. This ensures low latencies and better QoS. Others are investigating the use of adaptive forward error correction (FEC) protocols, application-specific filtering, and routing in wireless ad hoc networks using active networking techniques. Active ad hoc networks provide the opportunity for coexistence of different routing protocols operating in parallel, as well as the capability of wireless ad hoc nodes to switch from one routing protocol to another.

8.2 A Brief History of Wireless Networks

The evolution of wireless communication started with the invention of the wireless telegraph by Guglielmo Marconi in 1896. At that time he successfully communicated telegraphic signals across the Atlantic Ocean from Cornwall to St. John's, Newfoundland. He sent alphanumeric characters encoded in analog signals. In the middle of the 20th century, many key foundations of wireless communications were invented at the Bell Labs. At the beginning of the 1950s, the Bell Telephone Company in the United States introduced a telephone service for its customers. This was the first example of a telephony network for commercial use. However, this network was limited and could serve very few subscribers. As the demand for radio telephony service slowly grew, it forced designers to come up with better ideas to use the radio spectrum effectively to enhance capacity and serve more subscribers. The concept of shared resources was introduced in 1964.

In 1971, the Federal Communications Commission (FCC) in the United States allocated a frequency band for radio telephony. Bell Labs developed the cellular concept in the 1960s and 1970s. The advancements in the field of solid-state electronics gave a new dimension to wireless communications. The researchers at Bell Labs reused the limited radio frequency (RF) spectrum in a group of cells arranged to serve a number of customers. Additionally, calls are systematically handed over from one cell to another to accommodate users' mobility from one cell to another. These ideas and developments have revolutionized wireless communications and led to a multi-billion-dollar industry.

The Bell Telephone Company introduced the Advanced Mobile Phone Service (AMPS) radio network, thereby deploying the first cellular network. In 1982, the United States standardized the AMPS system specification, and this became the radio telephony standard for North America.

In the 1980s, many cellular radio networks were deployed around the world. In Europe, each country selected its own technology for analog cellular telephony. The U.K. and Italy chose the Total Access Cellular System (TACS). The Scandinavian countries and France deployed the Nordic Mobile Telephone (NMT) standard. Germany used the C-Net standard. All these were analog systems, and hence considered first-generation systems. The first generation of cellular networks was analog and used frequency division multiplexing (FDM) technology.

In 1982, the Conference of European Posts and Telecommunications (CEPT) established the Groupe Special Mobile, now known as Global Systems for Mobile (GSM), and proposed the creation of a European standard for mobile radio telecommunications. This group produced the GSM standard that is widely deployed today. It is based on digital radio telephony. GSM is also referred to as "2G" technology, or "second-generation." GSM is based on time division multiple access (TDMA) technology. In the United States, the Telecommunication Industry Association has developed two interim standards, the IS-54 standard in 1990, which is based on TDMA, and the IS-95 standard in 1993, which is based on CDMA.

Code division multiple access (CDMA) is a spread-spectrum air interface technology used in some digital cellular personal communications services and other wireless networks. The idea of spread spectrums was proposed during World War II in an effort to prevent the attempts of listening or jamming of radio signals. The two types of spread spectrum are frequency-hopping spread spectrum (FHSS), which uses rapidly alternating FM signals known in advance by the sender and receiver, and direct-sequence spread spectrum (DSSS), which covers a wide range of frequencies transmitted simultaneously. DSSS requires high bandwidth, usually in the megahertz (MHz) range.

The term *Bluetooth* refers to an open specification for a technology that enables short-range wireless communications. The Bluetooth Special Interest Group created version 1 spec in 1999, explicitly defining a method for wireless transports to replace serial cables that would be used with peripherals. Bluetooth wireless communication operates in the 2.4-GHz frequency range.

8.3 Current Trends of Research in Mobile and Wireless Networks

The wireless and mobile environment has created new areas of research in the field of networks in general and in the area of programmable networks in particular. Physical-layer faults (e.g., co-channel interference, hidden terminals, path loss, fast fading, and shadowing) lead to time-varying error characteristics and time-varying channel capacity.[3] The quality index maintained across the wireless channel is called wireless quality of service (QoS).[3] User mobility can cause rapid degradation in the quality of the delivered signal, resulting in handoff dropping. Because of this, mobile applications can experience unwarranted delays, packet losses, or loss of service. The quality index maintained during handoff between access points is called mobile QoS.[3]

There has been a lot of research in the field of wireless networks during the past ten years. The extraordinary growth in cellular telephony and wireless LANS over the past several years has led to an increase in demand for mobile multimedia services. This will require existing broadband and Internet technologies to be complemented by high-speed wireless services. There are many challenges for its implementation. Some of these challenges and their solutions are discussed below.

Multimedia services demand high bandwidth. This becomes a challenge for wireless networks. Bandwidth is the most critical resource in wireless networks, and thus, the available bandwidth of wireless networks should be managed in the most efficient manner. To increase the bandwidth effectively, small (micro, pico) networks are used. These networks face the problem of rapid handoffs among users due to smaller coverage areas of cells. This leads to higher dropping of connections and makes QoS guarantees difficult. Hence, one of the most important issues is how to reduce frequent handoff drops. Reserving high bandwidth for handoffs can reduce connection dropping in wireless networks. This may increase the chances of new connections being blocked and ultimately the low-bandwidth utilization. Hence, the number of dropped connections and the number of connections blocked cannot be reduced. Therefore, there is need for developing new and efficient bandwidth reservation schemes. Jau-Yang Chang et al. proposed probabilistic resource estimation by estimating switching times, the estimated staying times, the handoff probabilities, and the required bandwidths of mobile hosts of each base station. The amount of reserved bandwidth for each base station is dynamically adjusted.

M.E. Kounavis et al.[3] have discussed the concept of programmable handoffs to help the deployment of new services and mobility management needs of a wide range of devices. These handoff services are programmable and have the capability of working in changing environments. Kounavis et al.[3] have presented two types of programmable handoff techniques to address these issues. The multihandoff wireless network service can simultaneously support multiple types of handoff control over the same physical wireless access network infrastructure. The reflective handoff service dynamically injects signaling mechanisms into mobile hosts before handoff. With this technique, mobile devices can freely roam between heterogeneous wireless access networks that support different signaling systems for mobility management.

The design, implementation, and evaluation of a programmable architecture for handoff services need advances in software and networking technologies to cater to specific radio, mobility, and service quality requirements of the wireless Internet and other wireless network services.

Middleware is an important architectural system to support distributed applications. The role of middleware is to present a unified programming model to applications and to hide problems of heterogeneity and distribution. It provides a way to accommodate diverse strategies, offering access to a variety of systems and protocols at different levels of a protocol stack.[4]

The conventional middleware platforms offer a fixed programming model to applications, together with fixed per platform implementations. This is a very simple solution, as it hides the implementation details from the user. Recent developments in the areas of wireless multimedia and mobility require more openness and adaptivity within middleware platforms. Programmable techniques offer a feasible approach to avoid time-varying QoS impairments in wireless and mobile networking environments. Multimedia applications require open interfaces to extend systems to accommodate new protocols. They also need adaptivity to deal with varying levels of QoS from the underlying network. Mobility aggravates these problems by changing the level of connectivity drastically over time.

Geoff Coulson et al.[5] proposed the solution of adaptable middleware to counter these problems. Applications should be able to check internal components and also adapt the system at runtime to meet current application needs. They should be able to alter the behavior of particular components, select a different configuration of components to provide a particular service, or add new components to the middleware platform at runtime.[5]

The utility curves presented by Giuseppe Bianchi and Andrew T. Campbell[6] help to achieve QoS by capturing variations in the available bandwidth. During congestion, the packet scheduler in an adaptive QoS model may inspect each application's utility curve.

8.4 Handoff in Wireless Networks

Handoff is the process deployed to change the assignment of a mobile device from one base station to another as the mobile device or unit moves from one cell to another. Handoff is performed in different types of networks in different ways.

8.4.1 Handoff in Cellular Networks

This is dependent on a number of factors, e.g., relative signal strength, call-blocking probability, handoff-blocking probability, etc. The most important factor is the relative signal strength. The strength of the signal between the base station and the mobile unit must be strong enough to

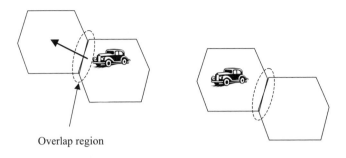

Overlap region

Figure 8.1   The handoff process in a cellular network.

maintain the signal reception quality. If the signal strength falls below a predefined threshold in the current cell, a handoff is initiated and the user registers itself with the new base station in a neighboring cell that has a stronger signal. Figure 8.1 shows the handoff process in a cellular network.

8.4.2 Handoff in Wireless Local Area Networks

The handoff process in a WLAN is shown in Figure 8.2. The access point broadcasts a beacon signal periodically (the period is around 100 ms). A mobile station (MS) that powers on scans the beacon signal. The beacon signal is a management frame that is transmitted periodically by the access point (AP). The beacon contains information corresponding to the AP, such as time stamp, beacon interval, and traffic indication map (TIM). The beacon is used to identify the AP. The MS uses the information in the beacon to differentiate between different APs.

The MS keeps track of the relative signal strength of the beacon of the AP with which it is associated, and when the relative signal strength becomes weak, it starts to scan for stronger beacons from neighboring APs. The MS chooses the AP with the strongest beacon or probe response and sends a reassociation request to the new AP. The reassociation request consists of information about the MS as well as the old AP. In response to this request, the new AP sends a reassociation response that has all the concerned information, e.g., supported bit rates, required to restart communication.

8.5 Active Base Stations and Nodes for Wireless Networks

There are two issues in mobile computing that must be overcome to implement QoS-aware communications. The first is the highly time-varying

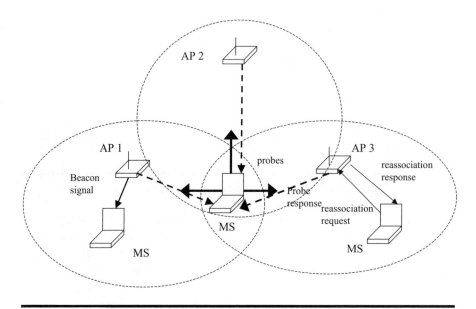

Figure 8.2   The handoff process in a WLAN.

nature of wireless links, particularly when location and speed are the main concerns. Second, the heterogeneous nature of an operational environment that is a mobile node, during the course of its movements, may encounter base stations that provide different services, protocols, and interfaces.

To deal with the heterogeneity problem, a base station should accommodate the mobile stations with differing capabilities that may visit it, and a mobile node must be able to operate with changes in available services and protocols as it moves to new base stations.

An important standard mechanism to deal with the variability and heterogeneity present in a mobile environment is adaptability. In fact, adaptability may be present in many different forms and at many different levels in a wireless network. At the lower levels, adaptability requires the communication and protocol processing to be dynamically modified, either by updating algorithm parameter values or, perhaps, by even changing the computation structure of the algorithm. For example, a base station may download an appropriate Medium Access Control (MAC) protocol when a new mobile enters its service area. Another example is that of a mobile device uploading a custom transcoding function at the base station to encode or scale an incoming multimedia packet stream according to the user interface capabilities of the mobile device.[2]

The approach adopted by A. Boulis et al.[2] allows users at the communication endpoints to program the base stations, and the endpoints to be programmed by the network. This approach allows application-specific customization and adaptation of packet processing at the base station and the end nodes at runtime.

As far as the adaptive communication processing and link protocols are concerned, adaptive equalization and adaptive error control are used to maintain robust performance in the presence of wireless link impairments, such as fading and multipath. Adaptability can also be used to improve energy efficiency. For example, A. Boulis et al.[2] describe a hybrid FEC-ARQ (automatic repeat request) error control method where, for each packet flow, a possibly different hybrid combination of an ARQ protocol and a FEC code is selected, depending on the quality of service (QoS) requirements, and the combination itself is adapted via QoS renegotiation as available effective link bandwidth changes.[2]

The architecture presented by A. Boulis et al.[2] consists of active base stations and active wireless nodes where application-specified custom processing can be done on packet flows. The packet processing is done using flow graphs consisting of packet processing function (PPF) entities. The PPFs are implemented in software (e.g., Java) or hardware through field-programmable gate arrays (FPGAs).[1] The hardware solution is faster than the software one.

Figure 8.3 explains this concept with the use of two scenarios. In the first example, an active mobile node enters into an insecure environment. Base station A (BS-A) is active, so the node can upload its own encryption algorithm in the form of a PPF. The particular PPF encrypts all packets intended for the node before they are sent to some other node. The active mobile node can change the encryption algorithm on the fly to ensure foolproof security. In the second example, there are two mobile nodes; each of these two mobile nodes uses a different style of handoff — one is soft and the other is hard. The handoff style is mobile initiated,

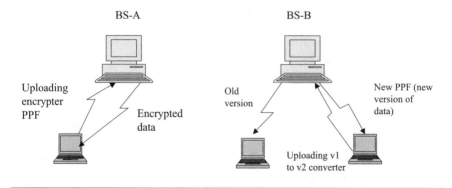

Figure 8.3 Uploading and downloading of PPF.

with the soft handoff on the downlink and the hard handoff on the uplink. By mobile initiated, we mean that some time after arriving at a cell, a mobile device initiates a handoff by first registering with the new base station, B (BS-B), or access point and downloading the relevant PPF. During soft handoff, the active mobile device or node simultaneously receives flows from the old and new access points on the downlink. In contrast, uplink flows use hard handoff, i.e., a terminate-and-establish approach between the old and new access points. The mobile device completely breaks connection with the old access point or old base station before connecting to the new access point or base station and resynchronizing itself to it.

One important thing to notice here is that apart from the flexibility one can get in the link management using adaptive techniques, the mobile nodes are no longer dependent upon the base station's administrator to provide them with new services. As long as the base stations are active and adaptive, the network engineers may simply develop PPFs to provide new services and protocols, thus facilitating the deployment of new technologies.

8.5.1 Architecture

The system consists of active base stations and active wireless nodes where application-specified custom processing can be done on packet flows. The packet processing is performed using flow graphs composed of packet processing function (PPF) entities. The PPFs are implemented in software (e.g., Java) or hardware, i.e., field-programmable gate arrays (FPGAs). A PPF can be anything, ranging from a programming language code fragment that is executed on the active node's central processing unit (CPU) to a bytecode used to program the FPGAs at runtime. PPFs are a part of a node, or it can be a function call to a particular application programming interface (API) to program the FPGA.

A PPF can take input from many different streams of data coming from other PPFs or the classifier, perform a collective function on them, and then produce many outputs, possibly different or the same, which can be passed to other PPFs.

All processing is done on packets while they move between the IP and radio after they are captured; i.e., incoming and outgoing packets at the network layer are available to the PPFs for processing. Only the packets that need to be processed by PPF flow graphs are captured. The application that receives or produces these packets is completely unaware of this processing. The process is completely transparent to the application that receives or produces these packets. With this process, an old application can be made functional in a new environment without the need

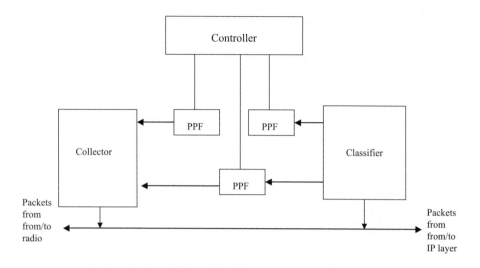

Figure 8.4   Packet flow through a network of PPFs. (With kind permission from Springer Science and Business Media.)

for modifications simply by installing an appropriate PPF flow graph somewhere along the packet flow path. As Figure 8.4 shows, the packets that are captured go through a network of PPFs.

The network accepts modifications by allowing new PPF flow graphs to be instantiated and old ones to be terminated. The instructions to install or terminate a flow graph begin at remote sites (a remote site is a mobile node or base station other than the node or base station that is currently executing the PPFs). These instructions consist of information regarding which packets to capture and process with a particular PPF flow graph, along with the description of the functions of what the PPFs will do. Although the remote site decides how the PPFs are set up, the actual processing of the packets is done locally on the node accepting the instructions and not at the remote sites.

The classifier receives the packets at the entrance to the PPF. Its function is to decide to which flow a packet belongs, and ultimately the PPF flow graphs that should process this packet. This decision is made by examining Transmission Control Protocol/Internet Protocol (TCP/IP) header information, for example, the destination and source IP addresses and TCP port numbers. At the other end of the PPF network is the collector, which collects processed packets from all the flows and serially writes them back to the capturing entity, where they can travel along their normal route to the radio or IP layer. The capturing entity is a driver written in C language that determines whether the packets should be captured and placed in a queue for later reading or continue along their route unhindered. This decision is made based on a list of destination and source addresses and ports updated by the controller. Packets are captured when a match of the addresses and ports is found.

This architecture provides active networks functionality because a PPF can be used in many PPF flow graphs and is therefore quite flexible. One can use previously uploaded code without the need to upload it again; it also implies that a node can share PPFs defined by other nodes. In this way, a mobile node has access to a large section of PPFs in addition to what it carries. These are PPFs that are predefined in the base station and PPFs that belong to other nodes that have allowed their use.

The controller communicates with remote sites to accept uploadable functions and function parameters. When the controller receives all information from the remote site (i.e., description of any new PPFs, how to connect PPFs in new or existing flow graphs, and what kind of packets to capture), it installs the new flow graph and asks the capturing mechanism to add new packet types to the capture list.[2] The controller can also end the execution of a PPF flow graph when instructed, or when PPF is not used for a considerable period. Other important functions that the controller performs are security monitoring, authentication, and PPF library maintenance.

8.5.2 Security

The controller implements security policy. It makes sure that a node can only install a PPF flow graph for packets that are destined for it or produced by it. The IP address determines the source of PPF instantiation commands. The IP address can be spoofed. This problem is solved by incorporating the authentication mechanism of IPsec.

Second, the code should not be able to access the memory of another application or PPF. This condition makes sure that malicious code should have the least effect on other applications (e.g., it should not be able to take over the resources of the system). This is a difficult problem to deal with when one considers a generic mobile code scheme. But fortunately, Java language provides a secure sandbox environment where the PPFs can execute.

Third, privacy has to be maintained. A user space application should not be able to read the buffer of the capturing entity. A signature with every call to the packet read function ensures this restriction.[2]

All these security issues become less severe in the hardware implementation of PPFs. There the resources are under better control, almost eliminating most security issues. A PPF will be assigned its own space on the FPGA, making it impossible to interact with other PPFs.[2] The only concern is the authentication that reconfigurations of the hardware are done properly. To ensure this, a tester will check the reconfiguration bit stream of a hardware PPF (that is, the equivalent of the mobile code for a software PPF) for hazardous reconfigurations. The hazardous reconfigurations may include two electrically connected outputs. Typically, such a checker is easy to write if appropriate details about the FPGA device are available.

The loader of a PPF defines a list of nodes that can have access to the PPF (i.e., will be able to instantiate it in their own PPF flow graphs). The interested node should acquire the key from the owner of the PPF. For its implementation, a secure key distribution mechanism is absolutely critical.

8.5.3 Resource Management

Because space on the reconfigurable hardware is limited, a resource management mechanism and admission policy for a new PPF flow graph are needed. To apply some admission policy, an active node should at least keep track of the free space available in the hardware, as well as the amount of hardware space used by each remote site. When a request to install a new hardware PPF arrives at an active node, the system should look at the amount occupied by the specific remote site and the total free hardware space remaining and make a decision based on these parameters. For example, if a particular remote site has used a large fraction of the hardware space, a new request would be denied, as we do not want a particular site to take over the hardware resources. In software, resource management gets more complex as instantiation of a new PPF affects the execution of other PPFs. Hence, a decision has to made about the admission of a new PPF flow graph by specifying simple QoS parameters, such as the maximum time for packet processing and minimum bandwidth usage of the PPF. The management of software PPFs becomes easy if the network operating system is a real-time operating system, because it makes sure that tasks are completed within the specific times.

8.5.4 Implementation

A. Boulis et al. used an off-the-shelf board with a Xilinx 6200 family runtime-reconfigurable FPGA chip. This particular FPGA has the advantage of *incremental reconfiguration*, as one can access the reconfiguration registers in random order. This kind of functionality is important if fast, on-the-fly, reconfigurable hardware functionality is required. In addition to the 6200 FPGA, the board has 1 Mbyte of synchronous RAM (SRAM) and a Xilinx 4013 FPGA that implements the peripheral component interconnect (PCI) interface and the driver's interface. After the design is correctly instantiated into hardware, input data is provided and the output data is produced.[2] This architecture deals effectively with the wireless link deficiencies and the heterogeneity of the mobile environment by using the active base stations and active mobile nodes.

Thus, PPFs provide new services and protocols, facilitating deployment of new services on a very short time scale.

8.6 Programmable Middleware Support for Adaptive Mobile Networks

To counter the time-varying QoS impairments of wireless mobile networks, O. Angin et al.[4] developed an open and programmable mobile network that is controlled by a software middleware toolkit called Mobiware. At the lowest level of programmability, Mobiware abstracts hardware devices and represents them as distributed computing objects based on Common Object Request Broker Architecture (CORBA) technology.[4] These objects (e.g., an access point object) can be programmed via a set of open programmable network interfaces to allow new adaptive services to be built using distributed object computing technology. Furthermore, the adaptive QoS algorithms are built as active transport objects based on Java objects and injected upon runtime into mobile devices, access points, and mobile-capable network switches/routers to provide value-added QoS support when and where needed.[4]

8.6.1 Protocol Stack of Mobiware

The Mobiware protocol stack consists of three layers: active wireless transport, programmable mobile network, and programmable MAC. It also consists of the Xbind broadband kernel. At the transport layer, an *active wireless transport* supports the end-to-end transmission of audio, video, and real-time data services based on an adaptive QoS paradigm. This transport layer functionality is implemented with a set of Java classes; the transport system binds active and static transport objects at mobile devices and access points to provide end-to-end transport adaptation services. Static transport objects consist of segmentation and reassembly, rate control, flow control, playout control, resource control, and buffer management objects. These objects are dynamically loaded into the mobile device as part of the transport service creation process to support value-added QoS. Examples of active transport objects are active media filters,[4] which perform temporal and spatial scaling for multiresolution video and audio flows, and adaptive forward error correction (FEC) filters,[4] which protect time-varying error characteristics.

The programmable mobile network layer supports the introduction of new mobile adaptive QoS services based on the Xbind broadband kernel.[4] The network layer supports switched IP flows over ATM native transport services. Architecturally, the network layer consists of a set of CORBA network objects and adaptation proxies that execute at the mobile device, access points, and mobile-capable switch/routers.

The three important Mobiware network algorithms are:

■ QoS-controlled handoff
■ Mobile soft state
■ Flow bundling

These algorithms are explained below.

The programmable MAC layer offers a programmable air interface to allow new services to be dynamically created and installed on the runtime. This is in contrast to supporting a specified set of hard-wired MAC services (e.g., constant bit rate) by means of a centralized control scheme. In fact, it adopts a distributed approach by transferring the application-specific adaptation decision making to mobile devices. The programmable MAC layer consists of a packet scheduler. This scheduler allows transmission of application flows on the basis of specified QoS requirements. During periods of congestion, the packet scheduler may check each application's utility curve.[4] A *utility curve* captures the adaptive nature over which an application can successfully adapt to available bandwidth in terms of a utility curve that represents the range of observed quality to bandwidth. The observed quality index refers to the level of satisfaction perceived by an application in terms of bandwidth at any moment.[4]

8.6.2 Summary of Programmable Objects

The set of programmable distributed CORBA objects supports the delivery of adaptive QoS flows to mobile devices. These objects execute on mobile devices, access points, and mobile-capable switch/routers supporting a set of mobile signaling and QoS adaptation algorithms (QoS-controlled handoff, flow bundling, and mobile soft state). The use of distributed object technology also provides support for interoperability between mobile devices using different operating systems and protocols.

The two per mobile (e.g., mobile devices, etc.) proxy objects that support programmable handoffs in Mobiware are QoS adaptation proxy (QAP) and routing anchor proxy (RAP).

8.6.2.1 QoS Adaptation Proxy (QAP) Objects

These objects allow mobile devices to probe and adapt to changing resource availability over the wireless link (e.g., availability of channel bandwidth).

8.6.2.2 Routing Anchor Proxy (RAP) Objects

These objects provide for mobility management services. They work with per mobile aggregation or bundling of flows to and from mobile devices for fast, efficient, and scalable handoff. The flow aggregation/bundling is discussed below.

The per mobile QAPs can be positioned at an access point or any mobile switch/router between the mobile and its corresponding RAP. If the QAP is present at the access point, the mobile soft state is only functional over the air interface, that is, between the mobile device and access point. The location of these proxies is programmable. Because bandwidth in wireless networks is generally the bottleneck, access points are the most suitable locations.

Mobiware uses a number of objects to dynamically control the functions of the network. Table 8.1 shows a summary of functions of different signaling objects. These signaling objects facilitate registration with the new access point, rerouting of flows, and QoS adaptation.

8.6.3 QoS-Controlled Handoff

Mobile device objects periodically look for beacon signals from neighboring access points. From time to time, the mobile device compares all beacons received over a current search period and cumulatively over multiple search periods. If the wireless QoS indicated in the beacon from the current access point falls below a predetermined threshold, the mobile device selects a new access point for handoff.

The device registration procedure starts the new access point object to bind to a mobility agent object. Mobility agents can operate at fixed-edge devices or mobile devices, or on the switches.

When the mobile device starts a handoff, it sends a unique mobile device identifier called the flow bundle identifier (FBI) to the access point, which allows Mobiware to identify the mobile device's flow bundle in the wireless access network.

Mobility agents are responsible for rerouting a mobile device's flow bundle from an old access point to a new one. Switch server objects are employed to reestablish a new flow state at all switches between the

Table 8.1 Summary of Functions of Different Signaling Objects

Object	Description
Mobile device object	Provides APIs for querying beaconing information, registering with new access points, establishing flows, renegotiating QoS, and handing off flows
Access point object	Supports APIs for binding wireless network objects to wireline network objects (e.g., mobility agent) on behalf of mobile stations; propagates CORBA calls and periodically refreshes wireless flow states; plays a vital role in QoS-controlled handoff and interacts with the transport system for the injection of active transport objects
Mobility agent object	Supplies adaptation and mobility management services; interacts with per mobile RAP and QAP states in the switch servers and supports APIs for retrieving network topology to establish, maintain, and hand off flows in the cellular access network
Switch server objects	Support APIs for the reservation and release of namespace, such as virtual channel identifier/virtual path identifier (VCI/VPI) pairs, and the allocation of network resources (e.g., bandwidth) in ATM switches

crossover switch and the new access point. The rerouting phase includes namespace reservation (outgoing VCI/VPI) and bandwidth value at each network switch and the new access point.

8.6.4 Flow Bundling

Flow bundling[4] provides a collective routing representation for all the flows to and from a mobile device. This is like the virtual path concept in ATM networks, tunneling in IP networks, or the DiffServ aggregates concept. Flow bundling reduces the signaling overhead. It also reduces the complexity of rerouting multiple independent flows to and from mobile devices during handoff. This speeds up the handoff process.

The mobile agent interacts with the switch servers to reestablish flows and update switch tables for all switches between the crossover switch and the mobile device. The General Switch Management Protocol (GSMP) is used to update the switch table at each traversed switch.[4]

8.6.5 Mobile Soft State

The soft state comprises information about the bandwidth and namespace resources for flow bundles exchanged between a mobile device and QAP. The mobile soft state provides a number of QoS adaptation support functions. For example, it helps in the rerouting of a flow bundle to a new access point during a handoff and in the reservation of resources. It also makes sure that the old flow bundle state between the old access point and the crossover switch is removed. This is achieved through mobile soft-state *timers* located at the switches and old access point, which time out and release resources automatically.

The soft state also supports mobile devices resident in cells in scaling flows in accordance with channel conditions, and when new mobile devices enter and leave cells. The network state is refreshed periodically through the soft state. The periodic refresh messages are sent by a mobile device as part of a soft-state probing mechanism on a per-flow-bundle basis. During the refresh phase, mobile devices react to any changes in allocated bandwidth (based on utility functions) for a flow bundle. The mobile devices issue a refresh to the mobility agent object, which then refreshes all switches and the current access point on the path between the mobile device and QAP with the current soft state. Mobile devices probe the path between the mobile device and the per mobile QAP from time to time and compete for resources.

8.7 Programmable Handoffs in Mobile Networks

8.7.1 Background

Handoff represents an important service in wireless networks characterizing the capability of access networks to respond to mobile user requirements and changing network conditions. Kounavis et al.[3] have presented a programmable architecture for profiling, composing, and deploying handoff services. This programmable handoff architecture uses collections of distributed algorithm modules that can be combined to compose different styles of programmable handoffs on demand. It also allows designing and dynamic deployment of handoff services based on number of users, radio, and environmental factors. The former process is called binding, and the latter, service creation environment. The architecture also employs code reuse, allowing different types of handoff services with simple software upgrades. A profiler is used for declaring programmable handoff services, the addition, deletion, and customization of programmable handoff objects, and the creation and removal of object bindings. Network operators can customize objects during the profiling process. In this case, parameters characterizing the operation of a service (e.g., user, service-specific, or environmental parameters) can be passed in objects at runtime through the profiler and service controllers.

The multihandoff service can simultaneously support three styles of handoffs over the same physical wireless access network that is commonly used in mobile networks: *network-controlled handoff* (NCHO), *mobile-assisted handoff* (MAHO), and *mobile-controlled handoff* (MCHO). This programmable approach can benefit the mobility management needs of a wide range of mobile devices, from cellular phones to more sophisticated palmtop and laptop computers, in a wireless Internet service provider's environment.[7] The other programmable handoff service is reflective handoff, which allows the dynamic injection of signaling system modules into mobile devices before handoff. With this service, mobile devices can continuously roam between different wireless access networks that support fundamentally different mobility management systems. What happens in reflective handoff is that each mobile device keeps a local cache of signaling system modules. Signaling system modules are basically collections of objects supporting mobility management services in mobile devices. Before a mobile device performs a handoff, it makes sure a signaling module associated with the new candidate access network is cached. If a signaling module is not present, it is dynamically loaded from the old access network through a two-way handshake. Access networks schedule the transmission of signaling modules over the air interface to avoid flooding of the wireless network. Timers associated with the modules are used to avoid large caches, and they are refreshed while a mobile device remains inside the coverage area of an access network with a set of modules. Figure 8.5 shows the programmable handoff architecture and the two processes: the binding process and the service creation process. The service creation environment offers interfaces in the form of a profiler for the dynamic introduction and modification of network services through transportable code. It also consists of a set of service controllers. The profiler works with a set of service controllers using a well-defined profile scripting language to create or modify programmable handoff services. Service controllers compile profiling scripts, resolve object bindings, and create handoff control and mobility management systems.

Programmable handoff services are implemented as collections of distributed objects. Interoperability between these different types of distributed objects is achieved through middleware technologies such as CORBA, Distributed Component Object Model (DCOM), etc. The programmable handoff objects expose control interfaces, allowing creation of bindings at runtime. In the binding process, an object obtains a reference to another object to request its services. An object reference can be a host name

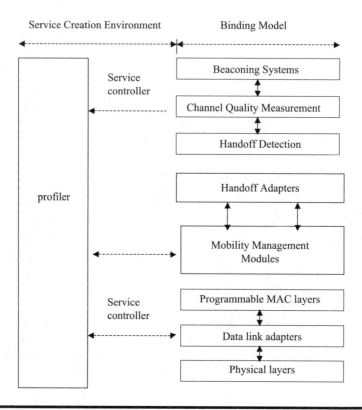

Figure 8.5   The programmable handoff architecture.[3] (With kind permission from Springer Science and Business Media.)

where the object is activated or a TCP/IP port number where the object listens for service requests.

The binding model consists of three models: handoff control, mobility management, and software radio. Each model has a separate service controller. The binding model supports separation of handoff control from mobility management, the decomposition of the handoff control process, and the programmability of the physical and data link layers. The handoff execution interface separates handoff control from mobility management. Handoff adapters integrate handoff control systems with mobility management services.

8.7.2 Handoff Control Model

A handoff control model divides the algorithms that support beaconing, channel quality measurement, and handoff detection. By separating the handoff detection and beaconing systems from wireless channel quality measurements, new detection algorithms (to be dynamically introduced, access networks and mobile devices) and different styles of handoffs (over the same wireless infrastructure) are supported, respectively.

The handoff control model supports the following services:

- **Detection algorithms**: Decide the most suitable access points (for example, base station in case of cellular networks) to which a mobile device should be attached. Wireless access points can be selected based on different factors, including channel quality measurements, resource availability, and user-specific policies.[8] A mobile device can be attached to one or more access points at any moment in time.[3]
- **Measurement systems**: Create and update handoff detection state. The handoff detection state is the data used by detection algorithms to make decisions about handoff.[3]
- **Beaconing systems**: Help in the process of measuring wireless channel quality. Programmable beacons can be customized to support service-specific protocols like QOS-aware beaconing[9] or reflective handoff.[3]

8.7.3 Mobility Management Model

The following services are identified as part of the handoff execution process:

- **Session rerouting mechanisms**: Rerouting services may include admission control and QOS adaptation for management of wireless bandwidth resources. These mechanisms control the routing paths in access networks to forward data to or from mobile devices through new points of attachment.[3]
- **Wireless transport objects**: These interact with the physical and data link layers in mobile devices and access points to transfer active sessions between different wireless channels.[3]
- **Mobile registration**: This is concerned with the state information a mobile device exchanges with an access network when moving from one point of attachment to another.
- **Mobility state**: This is based on the state of a mobile device's connectivity, addressing and routing information, bandwidth, namespace allocations, and user preferences.[3]

8.7.4 Software Radio Model

This model consists of the physical layer, data link adapters, and programmable MAC layer. Software radios permit mobile devices to dynamically tune to the appropriate air interface of the serving access network, while roaming between heterogeneous wireless environments. The most important part is the programmable MAC layer, which allows for the introduction of new adaptive services on demand. Data link adapters separate data link-layer modules from the lower physical-layer components. For example, data link adapters allow programmable MAC protocols to operate on top of different channel-coding or modulation schemes, as discussed in Kounavis et al.[10]

8.7.5 Handoff Adapters

A handoff adapter is an intermediate layer of distributed objects that joins the handoff control model with the mobility management model. Handoff adapters and mobility management services jointly implement handoff execution algorithms. Handoff adapters can be centralized (i.e., running in a single host or network node) or distributed. Distributed handoff adapters are used at mobile devices, access points, or mobile-capable routers/switches.

Handoff adapters activate mobility management services in a sequence that is specific to the handoff style being programmed. Mobility management services (i.e., session rerouting, wireless transport, mobile registration, and mobility state management services) are activated as part of the handoff execution process. For example, in a forward mobile-controlled handoff, an adapter would invoke a radio link transfer service before session rerouting. In the backward mobile-assisted handoff, the order of this execution would be reversed.[3] Each handoff style uses a separate adapter object. To activate mobility management services, adapters distribute method invocations to the network nodes or hosts where mobility management services are offered. Adapters hide the heterogeneity of mobility management architectures enabling intersystem handoffs. Adapters may interact with distributed mobility agents, databases supporting mobile registration information, or open network nodes to realize handoff in many different ways.[3]

This scheme improves network performance by offering programmable handoff services, but signaling overheads and scalability issues need to be investigated.

This programmable handoff architecture consists of two new services: multihandoff access network and reflective handoff services. A service creation environment can construct handoff services through profiling and composition techniques. The platform is capable of supporting multiple handoff control architectures over the same physical programmable access network by using a set of well-defined APIs and objects.

8.8 An Active Approach to Multicasting in Mobile Networks (AMTree)

8.8.1 Background

Multicasting is an important application to disseminate data to multiple clients. Applications that use multicasting range from videoconferencing to resource discovery. The IP multicast in the Internet is based on the Distance Vector Multicast Routing Protocol (DVMRP).[11] In this protocol, each group is identified by a group address and members come and leave as they want. Multicasting in heterogeneous environments is an area of research under investigation. Researchers are specifically looking at the routing of mobile hosts, because current multicast protocols like DVMRP,[11] Core Based Tree (CBT),[13] and Protocol Independent Multicast (PIM)[14] were designed with static hosts in consideration, and they suffer from following problems:

1. After migration, multicast protocols based on the shortest-path tree, such as DVMRP, may route packets incorrectly or drop packets due to reverse path forwarding.[11]

2. In approaches using a shared-tree mechanism such as PIM[14] or CBT,[13] an algorithm is needed to determine cores or rendezvous point (RP) strategic locations in the network. The cores are routers that provide abstraction during handoffs. That is, they make sure that no changes are required to downstream routers during handoff. Due to this strategy, receivers obtain the least possible delay. The core's location is usually selected at the start of the multicast session. In mobile networks, if sources and receivers are mobile, then a core's or RP's position becomes the least favorable after each receiver or source migration. So, relocation of the core on a dynamic basis is required in this case. Such relocation strategies should incur the least signaling overhead.

3. At the receiving site, when a mobile host (MH) migrates to a cell with no other group members, it will experience delay. This is mainly caused by subscription delay, tree rebuilding, or nonexistent multicast routers in the region.[15]

4. The time-to-live (TTL) field value specified may be inappropriate. For example, a TTL set for one region may not be suitable for

another. Once the MH gets out of a region, the earlier specified TTL value may be too small.

AMTree[15] handles these problems. AMTree is an active network-based approach for multicasting. It uses the source-rooted tree (SRT) approach.

8.8.2 The Problems of Mobile IP

8.8.2.1 The Tunnel Convergence Problem

Whenever the MH migrates to a new subnet, a bidirectional tunnel is created from the MH's care-of address to the HA. As a consequence, any traffic generated by the MH or directed toward the MH has to traverse through the home agent (HA). The tunnel is then used to send packets going to or from the MH. From the receiver's point of view, the source never left its subnet. This creates two problems: (1) a high handoff latency as the MH moves farther away from the HA and (2) the tunnel convergence problem. The latter problem becomes prominent when multiple MHs are serviced by one foreign agent (FA) and some or all have different HAs. As a consequence, each HA will have a tunnel to the FA. This leads to the tunnel convergence problem, creating bottlenecks of transmission.

8.8.2.2 Remote Subscription

When a mobile host migrates and is at a foreign network, it is assigned a care-of address. This address is then used for multicast. The drawback of this method is that if a source-rooted tree is used, then the tree needs to be rebuilt from the new location when the mobile host is at the foreign host. This incurs very high handoff latency because of the time taken to rebuild the tree. Due to mobility, packets directed toward the MH (such as negative acknowledgments (NACKs)) might not get through correctly and reconstruction of the tree each time the MH migrates is inefficient.

8.8.2.3 Receiver Migration

When a receiver migrates, three issues need to be addressed. First, the foreign subnet may not support multicast service. Therefore, the receiver is unable to rejoin the multicast session until it migrates to a subnet that supports multicast. Second, the foreign subnet may support multicast service but does not join the multicast group to which the visiting MH is subscribed. Third, in the case where the foreign subnet has joined the multicast group, the receiver may receive duplicate or subsequent packets.

8.8.3 AMTree

AMTree uses active routers (ARs) to provide abstraction during handoff. These active routers are referred to as cores. The cores are dynamic and distributed in nature. Hence, no traffic concentration is experienced. A core can be easily programmed, with monitoring and traffic management protocols applied only to parts of the tree. This is in contrast to DVMRP,[11] where one separate tree is constructed for each source; in AMTree, only one multicast tree is required for a given session.

The objective of AMTree is to enable the adaptation of the multicast tree during handoff. AMTree gives an efficient solution that requires no changes to the distribution tree after handoff. This enables a multicast source to be mobile while still being able to deliver data down the tree. Receivers have the option of efficiently using the tree when an alternative path to an optimal portion of the tree is found. Additionally, no periodic control messages are used to get topology changes. This is because receivers are required to join the tree explicitly; thus, data only flows over links that lead from the source to receivers. ARs dynamically filter out unnecessary control messages. For example, join/optimization/NACK/ACK messages are filtered out by the ARs closest to the subscribing receiver.

The AMTree protocol operates in three phases: (1) construction of active multicast tree, (2) update process during migration, and (3) tree optimization by active nodes.

8.8.3.1 Construction of the Multicast Tree

The construction of the multicast tree comprises three processes: join, leave, and send. A distributed location directory (DLD) service exists in the active network. Access to this directory service is achieved in the same way as it is achieved in the Domain Name Service (DNS).

8.8.3.1.1 Join Process

The join process is based on the following steps:

■ Like the traditional IP multicast protocol, a receiver indicates its interest in a multicast session to its access or local router. If an active session already exists, then there is no need to request the DLD for session establishment. If no active session exists, the local router asks the DLD for the contact point (source address). The contact point or the source address (or the router local to the MH) is updated whenever the source migrates.[15]

Table 8.2 States Maintained by Each AR

Data	Details
Upstream AR's IP	A link to the upstream AR is required to return any responses from receivers.
Subscriber's IP	The downstream ARs/end-hosts to multicast pockets.
Multicast address	The multicast address associated with the current session.
Time to live	The maximum duration in which the state is maintained.
Contact point	The root node of the multicast tree; is obtained from the LD or HOP_DISCOVERY message.
Forwarding IP	Used to tunnel packets to a recently migrated receiver.

Source: Kwan-Wu Chin and Mohan Kumar, *Mobile Networks and Applications*, 6, 361–376, 2001. (With kind permission from Springer Science and Business Media.)

■ A JOIN_REQUEST message is then sent hop by hop toward the contact point. Once the JOIN_REQUEST message is received by an AR that is subscribed to the session, a JOIN_ACK is sent back to the initiating router by the AR. This completes the join process.[15]

If a receiver is the first to be subscribed, then the processing AR loads the AMTree program and creates a state pertaining to the session.[15] Each AR is connected to an upstream AR and maintains a list of subscribers (which can be neighboring ARs or end hosts). Table 8.2 shows the states created at each AR for each session. Each state (parent and subscribers) is assigned a TTL value. The TTL value is refreshed after data is forwarded or received. When the TTL expires, for example, when the link leading to a downstream AR no longer exists, the corresponding states are removed.

In the multicast tree, ARs with at least one receiver are termed *core* routers. An example of a multicast tree with core and active routers is shown in Figure 8.6.

The main function of these core routers is to provide an abstraction during handoff. In other words, no modifications are required to routers that are downstream from the core router during handoff.

8.8.3.1.2 Leave Process

The leave operation in AMTree is performed explicitly. Any subscriber who does not want to receive packets sends a LEAVE_MESG to its corresponding AR. With this, the subscriber will be pruned from that AR. The subscriber also checks whether that AR has other subscribers. If no

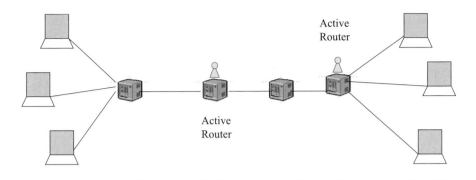

Figure 8.6 A multicast tree with core and active routers.

subscribers are found, then a prune message is delivered upstream. The upstream AR then removes the downstream AR from its list of subscribers.[15] AR has an associated TTL value. The TTL value is refreshed after packets pertaining to the session are processed. When a TTL value for a given subscriber has expired, its entry is removed.[15]

8.8.3.1.3 Send Process

The AR does not send any packets until there is at least one receiver subscribed to the session. Once the AR has a subscribed member (either ARs or receivers), it proceeds to multicast any packets sent by the source. When a packet arrives at an AR, the session details are accessed and a list of subscribers to the router is returned. The packet is then duplicated and forwarded to each subscriber.[15]

8.8.3.2 Handoff

The migration of the source involves two main processes: registration of the source's care-of address and connection to the nearest core. Once the connection to the BS is established, the handoff protocol executes the following steps:

1. At the start, registration of the MH's current care-of address is performed through the message REG_COF. This updates the source entry at the DLD with the MH's current care-of address. If an AR that has subscribed to the session is encountered by the REG_COF message, the intercepting AR sends a CORE-CONNECT to the MH's current location. The CORE-CONNECT message is utilized by the AMTree (loaded by AR) core discovery protocol to determine the

nearest node/core to connect to after handoff.[15] The CORE-CON-NECT message consists of four pieces of information: hop count, designated core, multicast address, and MH's care-of address.

2. A handoff update HO_UPDATE message containing the MH's care-of address is then sent to the previous contact point. A HOP_DISCOVERY with the MH's current care-of address is then generated by the AR and multicast along the tree (rooted at the old source). Cores on the tree receiving this message will then generate a CORE-CONNECT to the address specified in the HOP_DISCOVERY message. The HOP_DISCOVERY message continues to be forwarded along the tree until it reaches a leaf node where it is discarded.

3. A HO_COMPLETE message is sent back to the MH once the local AR has processed the CORE-CONNECT message.[15]

8.9   Advantages of AMTree

In AMTree, the multicast tree remains unchanged and is dynamically optimized. Additionally, ARs filter unnecessary control messages and additional customization of traffic can be easily added.

ARs are used to control duplicate NACKs and limit the delivery of repair packets to receivers experiencing loss. Different researchers have proposed different schemes using the AMTree concept. R. Wittmann and M. Zitterbart[16] have embedded quality of service (QoS) filters and error control mechanisms within active nodes. The active nodes thus transmit varying QoS streams depending on available bandwidth. In this case, different branches of the multicast tree have different qualities of service. Lau[17] worked on an active receiver-driven layered multicast protocol. This protocol does network filtering and achieves bandwidth convergence via the use of active routers.

AMTree closely follows the active multicast protocol proposed by Wetherall et al.,[18] which defines two capsules: *subscribe* and *multicast data*. Receivers joining the multicast tree send a subscribe capsule to the sender. The program in the subscribe capsule installs (or refreshes) forwarding pointers in the cache.[18] The multicast data capsule carries data and a program for routing. At each router, data is routed dynamically depending on the state information and program in the capsule. AMTree also uses the subscribe capsule to install state at routers.

The service provided by AMTree is connectionless, unreliable, and best effort. In other words, packets may be lost, duplicated, delayed, or delivered out of sequence. In this case, reliability is provided by higher layers like TCP/IP.

AMTree prevents the tunnel convergence problem because at any given time only one branch leads to a LAN with multiple MHs.

AMTree has low handoff latency, and interference to packet flow is negligible. This is made possible by using states and computation in the network that allow an MH to find a shortest path to the tree once it has migrated. As the tree undergoes nominal modifications, the MH can continue to multicast after migration. Therefore, the update resulting from migration is kept low, whereas in schemes such as remote subscription,[15] the entire tree needs to be rebuilt to accommodate the MH's new location.

The active multicasting protocol AMTree enables adaptation of the multicast distribution tree during handoff. It also helps the multicast tree to remain intact after handoff. This reduces end-to-end latency after migration of mobile hosts (MHs). The handoff latency has been shown to scale well as the number of receivers increases. In fact, the handoff latency remains fairly constant as the number of receivers increases. This is due to the higher probability of finding an AR subscribed to the session.

8.10 An Adaptive Management Architecture for Ad Hoc Networks

8.10.1 Background

In ad hoc networks, mobile nodes communicate via multihop wireless links. Ad hoc networks provide network connectivity without the need for fixed networking infrastructure. The inherent attributes of ad hoc networks, such as dynamic network topology, limited battery power, limited bandwidth, and a large number of heterogeneous nodes, make network management significantly more challenging than stationary and wired networks. In particular, the traditional management protocols fall short of addressing these issues.

C. Shen et al.[19] have presented the *Guerrilla management architecture* to provide adaptive and autonomous management of ad hoc networks. The management capability of Guerrilla is adaptive to accommodate the total number of nodes. It also accommodates heterogeneity of nodes.

The Guerrilla architecture has a two-level infrastructure to deploy management capability. The higher level consists of groups of *nomadic*, or roaming, managers that adapt to network dynamics, have more management intelligence, collaborate with one another, and serve as focal points for local nodes around them.[19] The lower level comprises lightweight *active probes* to perform localized management operations. Active probes are programmable codes that are made to be transmitted through agents or reasonably sized active packets. They are located inside managed nodes and remotely perform management tasks assigned by their nomadic manager to reduce bandwidth overhead. The other component of the architecture is the supervisor serving as a top-level manager. It controls and distributes management policies to a group of autonomous managers forming an agency that collaboratively carries out management operations. Each autonomous manager can adjust to its local environment, and hence is resilient to network dynamics.[19]

For scaling purposes, nodes are dynamically clustered into groups, with at least one nomadic manager in each group. The nomadic managers work together autonomously to manage the entire ad hoc network, with minimal help from external sources and the supervisor. Nodes are allowed to leave a group and join another when changing their locations. The corresponding nomadic managers cooperate to facilitate the handoff process. Due to the dynamic nature of ad hoc networks, the role of nodes serving as nomadic managers may change, depending on the topology, energy level, node density, and other attributes. For example, when the energy level of a nomadic manager falls below a certain threshold, the supervisor may decide to migrate to another capable node. When node density increases, a nomadic manager may decide to replicate its management functionality to another node that is capable of performing the role of a nomadic module to share the management load. Furthermore, when a nomadic manager anticipates network partitioning, it may decide to spawn another designated nomadic manager in the other partition to enable disjoint management operations. When two partitions reconnect, the corresponding nomadic managers may decide to unite.

To reduce management traffic and conserve wireless bandwidth, a nomadic manager may decide to issue an active probe (lightweight management code) to a distance-managed node. The active probe performs management tasks through interaction with the local Simple Network Management Protocol (SNMP) agent to achieve efficient use of limited wireless bandwidth by eliminating the need for constant polling of management information from remote SNMP agents. It also provides management operations when the quality of wireless links degrades or network partition occurs. Lastly, security mechanisms, such as trust, key distribution, and encryption algorithms, could be adaptively applied to accommodate the available network resources and the physical environment.

8.10.2 Node Classification

An ad hoc network might comprise heterogeneous nodes. These nodes will not be able to contribute equally to the management tasks. For instance, a sensor node may only have the capability to host an SNMP agent, while a switch or host has enough capacity to perform management intelligence tasks successfully.

Guerrilla architecture classifies nodes into three different roles of management capability according to their ability (e.g., energy level and processing power) and network conditions:

- Nodes playing the role of the minimum management capability only execute an SNMP agent to facilitate remote access to local management information in a client/server way.
- Nodes with sufficient ability are equipped with an active probe processing module, in addition to an SNMP agent. The active probe processing module is an execution environment capable of executing incoming active probes, which enables probes to not only query local SNMP agents to process information in a management information base (MIB), but also poll other remote SNMP agents. Active probes encapsulate lightweight management intelligence and perform assigned management tasks locally or collect local information from their managers.
- Nodes with enough energy levels and appropriate processing power can take up the role of a nomadic manager by executing the additional *nomadic management module* (NMM). A nomadic manager maintains management intelligence and states, collaborates with other nomadic managers, distributes active probes to other nodes within its management domain, and transfers or generates other nomadic managers according to network dynamics.

8.10.3 Active Probes

The function of active probes is similar to that of agents in SNMP. To get information from managed nodes and exchange management information with other nodes, active probes are employed to collect management information from these nodes. The active probes do not just merely collect raw data; they have the ability to locally process this raw data *inside* a node and selectively gather useful information before they report to the manager. Being active and adaptive, probes also collaborate with each other so that their information collection tasks can be performed even more effectively through filtering and aggregating techniques.

Active probes can be classified into two types based on their functions: monitoring and task specific.

8.10.3.1 Monitoring Probes

When a node changes its role to become a nomadic manager, it attempts to obtain surrounding network information by sending a monitoring probe to neighboring nodes. A monitoring probe may copy itself to cover the area specified by the nomadic manager, and collect and propagate local network information back to its manager.

8.10.3.2 Task-Specific Probes

To collect application-specific information or perform specific operations, such as quality of service (QoS) routing, probes can be specially programmed to perform these tasks. Normally, probes operate in a limited area and should not be programmed to go beyond a certain radius from the deploying manager. When a manager senses that the network covers a size larger than it can effectively manage, a new instance of the manager is spawned and sent to manage those distant nodes locally. The following features are required by active probes to efficiently serve the nomadic managers.

8.10.3.2.1 Independence

Monitoring probes are autonomous and do not require attention from the nomadic manager. They reduce the load on nomadic managers by freeing the managers from periodically polling other managed nodes.

8.10.3.2.2 Bandwidth Utilization

Monitoring probes are assigned to remote nodes and execute locally, which reduces bandwidth consumption incurred via conventional polling mechanisms.

8.10.3.2.3 Responsiveness

Network dynamics, such as movement of nodes, must be quickly reported to the manager so that information reflects the current network situation. Monitoring probes help to respond in real-time while filtering out unwanted updates.

Once a nomadic manager starts working, a monitoring probe is sent out, which quickly replicates itself to cover nodes within a specified radius from the nomadic manager. During this propagation process, a pointer to its parent node (where it is replicated from) is maintained. The pointer is used later when a probe returns any discovered information back to its nomadic manager.

In addition to announcing its presence to all the immediate neighbors, which is necessary for topology discovery, each node regularly broadcasts a *hello* (or *beacon*) message so that whoever is able to receive this message will realize that there is wireless connectivity between the node receiving the hello message and the node sending this message. In its first execution, the probe collects the node's local environment (obtained from the hello messages) and propagates this information back to its nomadic manager. This phase of operation is termed the exploring or location discovery phase.

In dynamic environments, the returning paths to the manager require a maintenance mechanism. In this architecture, immediately after a probe detects breakage of its upstream link, it sends a *route-to-manager failure notification* or *flush* message downstream. Upon receiving this flush message, a probe knows that it cannot contact the manager and starts looking for a new manager.

8.10.4 Nomadic Management Module

The Execution Environment for Nomadic Modules (EENM) serves as a virtual machine that hosts the NMM and facilitates intramodule communication. EENM also provides interfaces to other modules outside the NMM, such as the probe processing module. EENM waits for incoming management modules and performs necessary authentication and security checks. In the case of preinstalled management functionalities, these functionalities can be activated from another nomadic manager without code transfer to conserve bandwidth.

To provide adaptive autonomous deployment of management capacity within ad hoc networks, the NMM requires two features, discussed below.

8.10.4.1 Autonomy and Adaptiveness

The NMM possesses management intelligence, learns network dynamics from information collected via active probes, and adapts its behavior autonomously. It may also dispatch new probes and interact with other peer nomadic managers, based on information collected and management policy assigned by the supervisor.

8.10.4.2 Lightweight, Modular, and Extensible Design

The NMM must be lightweight enough to reduce migration or spawning overhead, and continue its management functionality at a new hosting node. In addition, modular design of nomadic management components will facilitate flexible and customized deployment. Also, the NMM functionality should be extensible to accommodate new or unforeseen requirements without interrupting the operations of the existing management capability.

In addition to supporting executions of the NMM, all management-related data is maintained inside a data structure, just like SNMP MIB. However, it is likely to be an aggregation of management information collected from neighbor nodes via probes. This collection of information is called the Guerrilla management information base (GMIB).

The Guerrilla management architecture thus employs a two-tier infrastructure to facilitate adaptive, autonomous, and robust management of ad hoc networks. The nomadic managers and active probes aid independent management operations and reduce consumption of wireless bandwidth. Nomadic manager nodes take care of information collection and adaptively perform management tasks when applicable.

8.11 Programmable Sensor Networks

Wireless ad hoc sensor networks (WASNs) have recently emerged as one of the key areas in wireless network research. Until now, these networks/systems have been designed with static and custom architectures for specific tasks, thus providing inflexible operation and interaction capabilities. Wireless ad hoc sensor networks are open to multiple transient users with dynamic needs. A. Boulis et al.[20] have presented a framework that supports lightweight and mobile control scripts that allow the computation, communication, and sensing resources at the sensor nodes to be efficiently harnessed in a need-specific way. The transport of such scripts in sensor nodes across a network allows dynamic deployment of distributed algorithms into the network.

These sensor networks are quite different from traditional networks. First, they have critical energy, computation, storage, and bandwidth constraints. Second, their overall usage scenario is quite different from that of traditional networks. There is not a mere exchange of data between users and nodes. The user will rarely be interested in the readings of one or two specific nodes, but instead will be interested in some parameters of a dynamic physical process. Moreover, the nodes that are involved in the process of providing the user with information are constantly changing as the physical phenomenon is changing. Therefore, the user interacts with the system as a whole.[20]

The SensorWare framework presented by A. Boulis et al. dynamically deploys distributed algorithms in several sensor nodes, which in turn

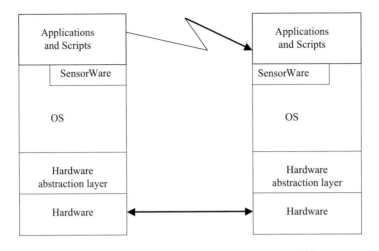

Figure 8.7 The architecture of a SensorWare.

means that these sensor nodes are dynamically programmed. A distributed algorithm is a set of collaborating programs executing in a set of nodes. The sensor nodes program other sensor nodes through these distributed algorithms.

8.11.1 Architectural Details

The architecture of SensorWare is organized into layers, as shown in Figure 8.7. The lower layers are the raw hardware and the hardware abstraction layer (i.e., the device drivers). An operating system (OS) is on top of the lower layers. The OS provides all the standard functions and services of a multithreaded environment that are required by the layers above it. The SensorWare layer provides the runtime environment for the control scripts by using the services offered by the OS layer. The control scripts rely completely on the SensorWare layer while populating around the network. Other services coexist with mobile scripts. They can utilize some of the functionality of SensorWare, as well as standard functions and services of the OS. These applications can be solutions to generic sensor node problems (e.g., location discovery) and can be distributed but not mobile. They will be part of the node's firmware. The SensorWare consists of (1) the language and (2) the supporting runtime environment.

A scripting language consists of functions/commands to be defined and implemented in order to use them as building blocks (i.e., these will be the basic commands of the scripts). Each of these commands will abstract a specific task of the sensor node, such as communication with other nodes or acquisition of sensing data. Furthermore, a scripting language needs constructs to tie these building blocks together in control scripts.

The basis tasks performed by the runtime environment are spawning of new scripts, scripts' admission control, and policing of resource use.

8.12 Summary

Active base stations and active wireless nodes facilitate application-specified custom processing on packet flows. The packet processing is done using flow graphs composed of packet processing function (PPF) entities. A PPF can be anything, ranging from a programming language code fragment executed on the active node's CPU to a bytecode that is used to program the FPGAs at runtime. A node can upload its customized encryption algorithm in the form of a PPF from a base station. The particular PPF encrypts all packets intended for the node before they are communicated to some other node. The active mobile node may change the encryption algorithm on the fly to ensure foolproof security. There is also a possibility of swapping the two styles of handoffs; one style is soft handoff and the other is hard handoff. The handoff style is mobile initiated, with soft handoff on the downlink and hard handoff on the uplink.

The programmable MAC layer in Mobiware offers a programmable air interface to allow new services to be dynamically created and installed at the runtime. Additionally, flow bundling provides a combined routing representation for all the flows to and from a mobile device. Flow bundling reduces the signaling overhead. It also reduces the complexity of rerouting of multiple independent flows to and from mobile devices during handoff. This functionality speeds up the handoff process. Mobility agents are responsible for rerouting of a mobile device's flow bundle from an old access point to a new one.

This programmable handoff architecture uses collections of distributed algorithm modules that can be combined to compose different styles of programmable handoffs on demand. It also allows designing and dynamic deployment of handoff services based on account user, radio, and environmental factors.

The multihandoff service can simultaneously support three styles of handoffs over the same physical wireless access network that is commonly used in mobile networks: network-controlled handoff (NCHO), mobile-assisted handoff (MAHO), and mobile-controlled handoff (MCHO). This programmable approach can benefit the mobility management needs of a wide range of mobile devices, from cellular phones to the more sophisticated palmtop and laptop computers, in a wireless Internet service provider environment.[15] The other programmable handoff service is reflective handoff service, which facilitates dynamic injection of signaling system modules into mobile devices before handoff. With this service, mobile devices can continuously roam between different wireless access networks that support fundamentally different mobility management systems.

AMTree filters unnecessary control messages, and additional customization of traffic can be easily added. AMTree limits NACKs implosion by controlling duplicate NACKs and limiting the delivery of repair packets. The active multicasting protocol AMTree enables adaptation of the multicast distribution tree during handoff. It also keeps the multicast tree intact after handoff. This reduces end-to-end latency after migration of mobile hosts (MHs).

The Guerrilla management architecture provides adaptive and autonomous management of ad hoc networks. The management capability of Guerrilla is adaptive to accommodate the total number of nodes. It also accommodates heterogeneity of nodes. Due to the dynamic nature of ad hoc networks, the role of nodes serving as nomadic managers may change, depending on the topology, energy level, node density, and other attributes.

SensorWave introduces programmability and opens networks to external users and systems through distributed proactive algorithms.

Exercises

1. What are the current research trends in wireless networks?
2. Explain the concept of programmable handoffs. Explain how PPF facilitates the handoff process in wireless networks?
3. Differentiate between multihandoff service and reflective handoff service.
4. What are the major advantages of using AMTree?
5. Discuss the role of nomadic managers and active probes in Guerrilla management architecture for ad hoc networks.
6. Discuss the architecture of SensorWave.

References

1. K. Pahlavan, *Principles of Wireless Networks: A Unified Approach*, 1st ed., Prentice Hall, Englewood Cliffs, NJ, 2002.
2. A. Boulis, P. Lettieri, and M. Srivastava, Active base stations and nodes for wireless networks, *Wireless Networks*, 37–49, Kluwer Academic Publishers, 2003.

3. M.E. Kounavis, A.T. Campbell, G. Ito, and G. Bianchi, Design, Implementation and evaluation of programmable handoff in mobile networks, *Mobile Networks Appl.*, 6, 443–461, Kluwer Academic Publishers, 2001.

4. O. Angin, A.T. Campbell, M.E. Kounavis, and R.R.-F. Liao, The Mobiware toolkit: programmable support for adaptive mobile networking, *IEEE Personal Commun.*, 5:4, 32–43, 1998.

5. G. Coulson, G.S. Blair, N. Davies, P. Robin, and T. Fitzpatrick, Supporting mobile multimedia applications through adaptive middleware, *IEEE J. Selected Areas Commun.*, 17:19, 1651–1659, 1999.

6. G. Bianchi and A.T. Campbell, A programmable MAC framework for utility-based adaptive quality of service support, *IEEE J. Selected Areas Commun.*, 18, 244–255, 2000.

7. A.T. Campbell, I. Katzela, K. Miki, and J. Vicente, Open signaling for ATM, Internet and mobile networks (OPENSIG'98), *ACM SIGCOMM Comput. Commun. Rev.*, 29:1, 97–108, 1998.

8. H.J. Wang, R.H. Katz, and J. Giese, Policy-Enabled Handoffs across Heterogeneous Wireless Networks, paper presented at the Second IEEE Workshop on Mobile Computing Systems and Applications (WMCSA'99), New Orleans, February 1999, pp. 51–60.

9. A.T. Campbell, M.E. Kounavis, and R.R.-F. Liao, Programmable Mobile Networks, *Computer Networks and ISDN Systems*, April 1999.

10. M.E. Kounavis, A.T. Campbell, G. Ito, and G. Bianchi, Accelerating Service Creation and Deployment in Mobile Networks, paper presented at the Third International Conference on Open Architectures and Network Programming (OPENARCH'00), Tel-Aviv, March 2000.

11. D. Waitzman, C. Patridge, and S. Deering, Distance Vector Multicast Routing, RFC 1075, November 1998.

12. J. Moy, Extension to OSPF, Internet Draft 1584, 1994.

13. T. Ballardie, P. Francis, and J. Crowcroft, Core based trees (CBT): an architecture for scalable inter-domain multicast routing, in *SIGCOMM'93*, San Francisco, 1993, pp. 85–95.

14. S. Deering, D. Estrin, D. Farinacci, V. Jacobson, C.-G. Liu, and L. Wei, An architecture for wide-area multicast routing, in *SIGCOMM'94*, August 1994, pp. 126–135.

15. K.-W. Chin and M. Kumar, AMTree: an active approach to multicasting in mobile networks, *Mobile Networks Appl.*, 6, 361–376, Kluwer Academic Publishers, 2001.

16. R. Wittmann and M. Zitterbart, Amnet: Active Multicasting Network, paper presented at the Proceedings of the International Conference on Communications (ICC'98), 1998.

17. W. Lau, On Active Networks and the Receiver Heterogeneity Problem in Multicast Session, honors thesis, Curtin University of Technology, Western Australia, 1998.

18. D.J. Wetherall, J. Guttag, and D.L. Tennenhouse, ANTS: A Toolkit for Building and Dynamically Deploying Network Protocols, paper presented at IEEE OPENARCH'98, San Francisco, April 1998.

19. C.-C. Shen, C. Srisathapornphat, and C. Jaikaeo, An adaptive management architecture for ad hoc networks, *IEEE Commun. Mag.*, 41:2, 08–115, 2003.
20. A. Boulis, C. Han, and M.B. Srivastava., Design and Implementation of a Framework for Efficient and Programmable Sensor Networks, paper presented at MobiSys 2003: The First International Conference on Mobile Systems, Application and Services, San Francisco, May 2003.

Chapter 9

Security in Active and Programmable Networks

9.1 Introduction

The term *Internet* brought a tremendous revolution in this modern age. The world became a global village, allowing anybody to access information from any place. The Internet is basically a collection of networks that may be contradictory and are joined collectively by means of gateways that control data transfer and transformation of communication from the conveyance networks' protocols to those of the receiving network. The Internet is a backbone of world communication between people, major nodes, or host computers, consisting of thousands of commercial, governmental, educational, and other organizations.

9.1.1 General Security Issues of Networks

The tremendous rate of use of the Internet over the last 15 years has created opportunities for information creation and sharing, but it has also created security risks for Web servers, data stores, the local area networks that host Web sites, network equipment, and even innocent users of Web browsers. Web security threats are classified as passive and active. Passive threats consist of eavesdropping on network traffic between the browser and server and accessing information on a Web site that is considered to be restricted. Active threats comprise attacks of impersonation, alternation of messages in transition between client and server, and the changing of information on a Web site.

These threats are most severe from the Web master's point of view. When a Web server is installed for an organization, the local area network (LAN) of that organization gets exposed to the entire Internet. Most Web site visitors browse through its contents for information, but a few will try to peek at things they are not supposed to look at. Others, not content with looking, will even try to change the information. The results can range from merely embarrassing, for instance, the discovery that an organization's home page has been replaced by obscene material, to damaging, for example, the stealing of a company's entire database of customer information. There are a number of approaches to ascertain Web security. These approaches are quite similar in their operations and differ to some extent in the mechanisms they utilize and with respect to their relative location in the Transmission Control Protocol/Internet Protocol (TCP/IP) stack. For example, one approach to implement Web security is to use IP security. It is transparent to end-user applications, and it also includes a filtering ability, so that only selected traffic constitutes the overhead for processing. The other solution is to implement security at the Secure Sockets Layer (SSL) level.

From the network administrator's point of view, the general objective of network security is to keep away strangers. However, the aim of a Web site is to provide the outside world with restricted access to information. Drawing a line and creating a division can be difficult. An unsuccessfully configured Web server can strike a crack in the most carefully designed firewall system. An unsuccessfully configured firewall can render a Web site open to attacks and impossible to use. Things get particularly complicated in an intranet environment, where the Web server must usually be configured to distinguish and confirm various groups of users, each with separate access privileges.

Cryptography is at the core of network security. It is a procedure in which a message is encrypted at the sender site and afterwards decrypted at the receiver end. In cryptography, the encryption/decryption algorithms are public, but the keys are kept secret.

In symmetric key cryptography, the same key is used by the sender (for encryption) and the receiver (for decryption); the key is shared. Symmetric key cryptography is usually utilized for long messages. Public key cryptography uses two keys: one public key and one private key. Public key algorithms are more efficient for short messages, because the key size is large.

Security is still an open issue in the area of active networking and in other relevant areas of research, such as mobile software agents. Even if secure paradigms that conform to the appropriate security and safety requirements are proposed, these paradigms should be tested in large-scale networks before final conclusions are drawn regarding their working and actual security.

The current security research on active networks can be classified into two general groups. The first deals with the more traditional view of security. It comprises authentication, access control, policies, and enforcement. Some examples include the protection of valuable information using encryption and the provision of data integrity using signatures. Public key infrastructure (PKI) and key distribution and management problems fall into this category.

The second group is based on the security associated with the mobile nature of the environment. Protection of nodes from mobile code originating in foreign domains and protection of active packets or code from malicious hosts both fall into this category. Protection from mobile code is usually provided by the security features of the language used as the execution environment.

There has been little research on dynamic, flexible, and application-specific security features that can work with the available dynamic functionality provided by active networks. Like traditional networks, active networks depend heavily on the underlying operating system for network security. Secure operating systems offer a solid foundation for application-level security services. The increases in networking and distributed mobile computing make the security support of operating systems even more critical.

Current active network operating systems do not have explicit security support. Applications cannot flexibly specify and enforce security and protection requirements according to their needs. Although a wide range of active policy types[16] and systems[17] have been proposed, underlying operating systems implement only a static subset of these policies and mechanisms. The overhead associated with adding new policies and mechanisms can also be counterproductive in some situations. The inflexibility of systems makes security policy and service customization complex and often leads to security gaps.

This chapter describes in detail the general security threats in active and programmable networks. It then discusses the different architectures to ensure security for active and programmable networks.

9.1.2 Types of Security Risks in Networks

There are basically three overlapping types of risks:

- Bugs or configuration problems in the Web server that allow unauthorized remote users to:

- Steal confidential documents
- Execute commands on the server host machine, allowing them to alter the system
- Gain information about the Web server's host machine that will allow them to break into the system
- Launch denial-of-service attacks, rendering the machine temporarily unusable
- Browser-side risks, including:
  - Active content that crashes the browser, damages the user's system, breaches the user's privacy, or merely creates an annoyance
  - The misuse of personal information knowingly or unknowingly provided by the end user
- Interception of network data sent from browser to server or vice versa via network eavesdropping. Eavesdroppers can operate from any point on the pathway between browser and server, including:
  - The network on the browser's side of the connection
  - The network on the server's side of the connection (including intranets)
  - The end user's Internet service provider (ISP)
  - The server's ISP
  - Either ISP's regional access provider

It is important to realize that secure browsers and servers are designed to protect confidential information against network eavesdropping. Without system security on browser and server sides, confidential documents are susceptible to interception.

Wireless LANs and wide area wireless networks are as popular as wired networks. WLANs provide opportunities to users to enter and leave networking domains frequently. This provides flexibility, but it also makes the system more vulnerable to security attacks. Hackers have found wireless networks relatively easy to break into, and network administrators must be aware of these risks and stay up to date on any new risks that arise. Also, users of wireless equipment must be aware of these risks, so as to take personal protective measures. Currently, a great number of security risks are associated with wireless technology. Some issues are more critical than others.

9.2 Types of Threats to Wireless Networks

There are many different styles and advanced ways through which unauthorized access can be made to an organization's wired and wireless networks.

9.2.1 Accidental Association

When a user turns on his laptop and it latches on to a wireless access point of a neighboring organization or company's overlapping network, the user may not even know what has occurred. This may result in the stealing or exposing of secret information of the organization through a link created between the two organizations.

9.2.2 Malicious Association

In this type of attack hackers connect to a company network when they access the network nodes through their hacking laptop instead of the company access point. These laptops (soft access points) are created when a hacker runs specific software to access the wireless network card. Then hackers are capable of stealing passwords, installing trojans, and instigating or initiating attacks. Wireless 802.1x authentications provide security, but it is still insecure to hacking.

9.2.3 Ad Hoc Networks

Ad hoc networks are peer-to-peer networks between two wireless computers that have no access point between them. These networks are exposed to greater security risks, such as malicious attacks from hackers and eavesdropping.

9.2.4 Man-in-the-Middle Attacks

These hackers are intelligent and connect to an access point through a wireless card offering a steady flow of traffic through the transparent hacking computer to the real network. The hacker can then sniff the traffic for usernames, passwords, credit card numbers, etc.

9.2.5 Denial of Service

A hacker or attacker continuously hits a specific access point or Web site with bogus requests, hasty successful connection messages, failure messages, and other commands. This may load the network and deny access to genuine users, eventually bringing down the network.

9.2.6 Network Injection

The hacker brings in bogus networking reconfiguration commands that affect routers, switches, and intelligent hubs. This brings down the network, requiring rebooting and reprogramming of networking devices.

9.2.7 Identity Theft (MAC Spoofing)

Most wireless systems allow some kind of Medium Access Control (MAC) filtering to permit only authorized users with specific MAC addresses to gain access and use the network. A hacker can listen to the network traffic and identify the MAC address of a computer with network privileges.

9.3 Security and Safety Issues of Programmable/ Active Networks

9.3.1 Difference between Security and Safety

Security and safety are two reliability properties of a system. *Safety* provides protection against errors of trusted users, meaning reducing the risk of mistakes or unintended behavior. *Security* means protecting against errors introduced by untrusted users, i.e., the process of protecting privacy, integrity, and availability in the face of malicious attack.

9.3.2 Main Threats to Active/Programmable Networks

As active networks are much more flexible than passive networks, the security threats are fairly high. A single packet that carries executable code can potentially change the state of a node. Hence, the security requirements for the computational environments where this code will be executed must be very strict.

In an active network environment, active packets may misuse active node resources and other active packets in various ways. There are instances where active nodes may misuse active packets. Some of the problems that may arise are discussed below.

9.3.2.1 Damage

Active network packets may damage or alter resources of a node by reconfiguring, erasing, or modifying them. If active packets share the same environment, they may attack each other. A node, on the other hand, may delete an active packet before completion of its function.

9.3.2.2 Denial of Service (DoS)

This problem has already been discussed above in the context of wireless networks. An active packet may overload a node service by constantly consuming its resources or using a great portion of the central processing unit (CPU) cycles. This attack degrades network performance.

9.3.2.3 Theft

This threat corresponds to the stealing of private information from a node through active packets. An active packet is also vulnerable to misuse when it visits a node because the information it carries has to be decrypted at the time of execution.

9.3.2.4 Compound Attack

In this type of attack malicious users send many active packets to a router in an uncontrolled way. This results in overflow of routers' queues and packets are discarded.

9.3.3 Protection Techniques

Some general techniques are presented that address the above-mentioned problems.

9.3.3.1 Authentication of Active Packets

Public key signatures can be used for the authentication of active packets, but there is no guarantee that active packets will be harmless.

9.3.3.2 Monitoring and Control

A reference monitor restricts the information, resources, and services of the system that active packets are permitted to use. It uses a security policy to determine if access is to be granted. The decision of granting permission to active packets through some credentials means that public key signatures are used.

9.3.3.3 Limitation Techniques

This technique decides the time period for which active packets are expected to be executed. Range limits for traverse and duplication are necessary for protection from monopolizing node resources.

9.3.3.4 Proof-Carrying Code (PCC)

PCC is based on the inspection of the code. The active node may simply confirm the proof and then run the program. The tricky part is the design of the proof, but this is the job of the program designer.

9.3.4 Protecting Packets

Two methods are recommended for the security of active packets: encryption and fault tolerance techniques.

9.3.4.1 Encryption Technique

This technique is used when active packets have no clear text code and data. Clear text data means meaningfulness of data for the receiver without deciphering.

9.3.4.2 Fault Tolerance Techniques

Fault tolerance techniques are of three types: replication, persistence, and redirection.

- Replication means that packets are reproduced at each node.
- Persistence means that packets are stored in a node for a short time, so if a node crashes, the copy persists in storage.
- Redirection means that packets may take other routes in case their default route fails.

Persistence and replication are rarely used because of the consumption of memory and bandwidth; it is used only when the need is very critical, e.g., packets installing the latest version of a routing protocol in all nodes. Although encryption and redirection seem to be good choices, they consume CPU cycles. These are basically immature techniques, so to achieve required results, there is a need for some extra work. When an active packet carrying data arrives at a node, the system should:

- Recognize the authenticity of the packet's credentials
- Identify the sending network element
- Verify the sending user
- Allow access to appropriate resources based on these identifications and credentials
- Permit execution on the basis of the authorization and security policy
- Manage and monitor all executions in a proper way
- When necessary, encrypt the packet and secure its code and data

If the active packet is not recognized in a proper way, then execute it in a constrained atmosphere or do not execute the code at all.

9.4 Active Network Security Research Projects

Active and programmable security has been the focus of many research groups around the globe. There are various groups that are engaged in research in this area. The nodes in active and programmable networks are exposed to external codes through active packets or agents to change their functionality. This increases the risk of damaging resources of nodes. This section describes the following active networks' security projects:

- Secure active network environment (SANE)[1,2]
- Safetynet[3]
- Secure Active Network Transport System (SANTS)[4]
- Secure QoS handling (SQoSH)[5]
- Programming Language for Active Networks (PLAN)[6]
- Active edge tagging (ACT)[7]
- Active security support for active networks[8]
- Dynamic access control mechanism[9]

9.4.1 Secure Active Network Environment (SANE)

9.4.1.1 Background

The Switchware[1,2] active network architecture consists of three tiers: active packets, which consist of mobile programs that replace traditional packets; active extensions, which are active codes that change the functionality of network elements and can be dynamically loaded; and a secure bootstrap architecture, which provides a high-integrity base on which the security of the other layers depends. Figure 9.1 shows the Switchware architecture.

These layers facilitate deployment of a variety of different approaches to meet the challenge of providing security in a programmable network, while still maintaining the flexibility of programmability and leaving the network usable.

In Switchware, active packets carry programs consisting of both code and data, like capsules in the Active Network Transfer System (ANTS). Basic data transport can be implemented with code that takes the destination address part of its data, looks up the next hop in a routing table, and then forwards the entire packet to the next hop. At the destination, the code delivers the payload part of the data. Traditional headers and trailers are used for conventional routing.

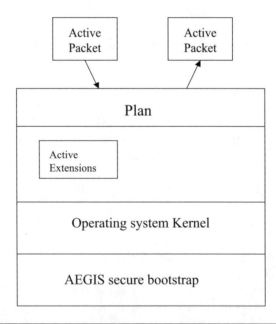

Figure 9.1 The Switchware architecture.

The active extensions are active codes that change the functionality of network elements. They use active packets to communicate with other routers and they themselves are static. They can be written in general-purpose programming languages, although type safety still plays an important role. They can use a variety of security mechanisms, including cryptography-based authentication and program verification. The AEGIS secure bootstrap layer, which deals with integrity aspects, will be discussed in Section 9.4.14.

9.4.1.2 Architecture of SANE

The SANE architecture is based on the following design principles:[2]

- ■ Dynamic checks are performed while the system is in use.
- ■ Static checks are performed before the system enters the operating state. These can be more expensive because they are done only once.

The static checks are very secure but can be costly, and the dynamic checks should be cheap enough so as not to degrade performance. Thus, there is a trade-off between cost and level of security.

Integrity and trust relationships in active networks are of critical importance. From the Open System Interconnect (OSI) model perspective, in a layered architecture, each layer in a system trusts the layer below it.

For an active network node, trusted node architecture can be developed by making the lowest layers of the system trusted, and then ensuring that higher layers depend on the integrity of these lower layers. There is, however, a problem when the actions of the node are programmable and dynamic and the programs are loaded from hosts or other active network elements. In this case, trust is not enough: a downloaded program from a trusted node may be corrupted and may damage the receiving node. *Dynamic integrity* checks ensure that the node remains a participating element of the active network in spite of such threats.[2]

The other approach is to limit the programmer in the choice of language and, within the context of this language environment, to restrict programs to those that are potentially secure. The technical advantage of this approach is that many security properties (e.g., access to regions of memory) can be analyzed at compile time, and thus checked once when compilation takes place, rather than dynamically at runtime. Thus, this design approach can provide security-based restrictions on program actions while preserving good performance.

9.4.1.3 Public Key Infrastructure

A very important element of this approach is the public key infrastructure. It is assumed that every user (or group of users) and every active element owns a public/private key pair, and that these keys (and certificates) are used to verify and authorize actions of those entities.[2]

A balance has to be maintained between flexible access and resource control policies. Finally, depending on the underlying network fabric, a certificate is revoked through expiration; this minimizes network traffic when authorization checks are performed.

The lower layers of the architecture ensure that the system starts in a required correct state. This is achieved by using a secure bootstrap architecture called AEGIS. This is a static integrity check, and after that, dynamic integrity checks are performed on a per user or per packet basis. These integrity checks are performed with a digital signature. The higher layers of the architecture are responsible for these checks. The system maintains security in several ways from this point onward:[2]

- First, it performs remote authentication, when required, for node-to-node authentication.
- Second, it provides a restricted execution environment for evaluation of the programs received by the network.
- Finally, it uses a naming scheme to partition the node's service namespace between users.

9.4.1.4 AEGIS Layered Boot and Recovery Process

In the AEGIS boot process,[2] either the active network element is started in the initial state, or a recovery process is entered to repair any integrity failure identified. Once the repair is completed, the system is restarted to ensure that the system boots. This entire process occurs without user interference. The boot process in AEGIS has been segregated into several levels to simplify and organize the AEGIS BIOS modifications, as shown in Figure 9.2. Each increasing level adds functionality to the system, providing correspondingly higher levels of abstraction. The lowest level is level 0. Level 0 contains the small section of *trusted* software, digital signatures, public key certificates, and recovery code. The first level comprises the remainder of the usual BIOS code and the complementary metal-oxide semiconductor (CMOS). The second level maintains all of the expansion cards and their associated read-only memories (ROMs), if any. The third level consists of the operating system boot blocks, which are responsible for loading the operating system kernel. The fourth level comprises

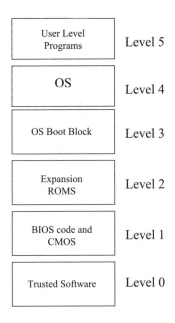

Figure 9.2   The boot process in AEGIS.

the operating system, and the fifth and the last level contains user-level programs and any network hosts.

The transition between levels in a traditional boot process is accomplished with a jump or call instruction, like a typical remote procedure call (RPC) with no security concerns. AEGIS employs public key cryptography and cryptographic hashes to protect the transition from each lower level to the next higher one, and in the event of a failure, a recovery process is initiated by using a trusted repository.

In the recovery process, a node that has detected an integrity failure can establish a secure channel with a repository. It then requests a new version of the failed component from the repository. The failed component can be any software or a piece of hardware. The repository will send the new component protected by the shared key or digital signatures, to prevent tampering from an intruder.

The detection of an integrity failure causes the system to boot into a recovery kernel contained on the network card ROM. The recovery kernel contacts a trusted host to recover a signed copy of the failed component. The failed component is then replaced, and the system is restarted.

9.4.1.5 Dynamic Resource Naming

Resources such as memory, CPU cycles, bandwidth, disk space, and real-time guarantees will generally be shared between different sessions of the same user, or even between different users. These users will need to identify (name) the particular resource they want to use.

The method of using some user-defined value would not work well, because names need to be unique across the active network. If users arbitrarily assign names to their resources, it is possible that there will be accidental naming conflicts, and there is a possibility of faking the names as well. Alternatively, some centralized authority could assign names per request, making sure these remain unique; this solution is unattractive because it does not scale well as the number of names required increases.

SANE has presented a decentralized way of naming dynamic resources that does not allow name collisions or accidental or malicious attacks. It works on the assumption that to load a module on the active element, the user (owner of the key) must pass some type of authorization check.[2] Furthermore, this authorization is fine grained; each user or group of users is unique. This assumption is reasonable, because we expect that an active element owner will probably want to limit the resources any user would potentially consume. Additionally, the network owner will want to give different access and resource rights to different users.

Hence, Switchware provides a range of different flexibility, safety and security, performance, and usability trade-offs by using a layered approach.

9.4.2 Safetynet

The Safetynet project at the University of Sussex deals with the issue of designing a programming language for active networks.[3] By mixing the semantics of computation with a pragmatic description of the processes of packet communication, researchers have produced a programming model that protects the connectivity of the network and places specific requirements to ensure security, safety, and fair resource allocation.

The design of the programming model is based on three major requirements: communication, safety and security, and programming. Some of these requirements are as follows:

- **Communication requirements**: The first characteristic of current routing active networks protocols is that given a destination, a program should be able to discover the next hop on the route to the destination. The second requirement is that the next active network hop should be transparent to the actual number of active-network-unaware nodes in between the two active network nodes. The third requirement is that the state installed in a node must be reinstallable or valid when reappearing after a temporary absence. This is critical because nodes in a network are subject to arbitrary reboots and disconnections. In the context of active networks, their state should not be lost, but saved. Finally, the network condition to any next hop should be available. This requirement is needed so that decisions on possible courses of action are based on network characteristic estimates.[3]

■ **Bandwidth requirements**: These are based on restrictions on the number of packets entering the network. The number of packets generated per unit time from a single node must also be bounded. These two requirements make sure that a program cannot generate a denial-of-service attack as far as bandwidth is concerned.

■ **Processing time requirements**: Packet-forwarding code must not be able to enter an infinite loop, and it should be executed as fast as possible. There must be a bound on the amount of processing time per unit time a thread can consume.

■ **Memory requirements**: Because each thread takes up memory, there should be a bound on the number of threads a program can generate. Also, there should be a bound on the amount of memory a program running on a node can consume. Another requirement is that programs cannot generate random references to memory. Because heap allocation takes time, it is not desirable for packet-forwarding code to allocate heap. Finally, an active network program should not directly manipulate the routing table of an active node. This requirement avoids disconnections in the network but restricts the power of active packets. However, it is much safer for active packets to indirectly change the routing table by calling relevant routines from the routing protocol, instead of directly maneuvering its contents.[3]

■ **Security requirements**: The security model of Safetynet is based on a set of trusted nodes and trusted code that is protected using cryptographic techniques. It must be possible to trace a chain of trust from a given code to a trusted node. Also, this chain must not be forgeable.[3]

■ **Programming requirements**: The use of a strongly typed programming environment that embodies policies about safety and security within the type system statically proves that a program is safe. Therefore, a type system that supports the safety and security concerns of the model is a requirement as well. This works with static checks, and they are more complex than runtime checks. A language that matches with all the above requirements may not be enough to provide security in an active network environment. However, it will remove the costly runtime checks that many current languages have.[3]

9.4.3 Secure Active Network Transport System (SANTS)

SANTS provides for the inclusion of authorization information in the active packet itself, so that the packet can cross multiple administrative domains and still be properly monitored.[4] This security architecture ensures enforcement of each individual node's authorization policy for access to its services and resources, providing the nodes with the assurance that their assets can be protected.[4]

Through global credential identifiers and a ubiquitous policy language, SANTS provides for flexible end-user control over authorization of access to its created state in the network. This ensures that the end user's data and service will be protected according to his or her needs. It meets the basic requirement that the node data will be protected even in the face of relaxed policies of the code downloaded.

SANTS is based on ANTS, developed at MIT.[4] It created an execution environment supporting strong security by extending the MIT ANTS execution environment.[10] SANTS adds the following capabilities to ANTS:

- X.509v3 certificates
- Domain Name Service (DNS) CERT records for storage of the credentials
- Java Crypto application programming interface (API) and Java Cryptographic Extensions Crypto provider
- Keynote policy system
- Java 2 security features
- A separation between execution environment (EE) and node classes
- A shared data capability, called BulletinBoard

9.4.3.1 Authentication Process

Certificates are computer-based files or structures used to send information about a user for identification purposes. They are based on International Telecommunications Union (ITU-T) Recommendation X.509.[11] A certificate binds an identity (e.g., a name) to a public key. The certificate comprises the name of the individual (e.g., John), the individual's public key, and a digital signature for the data. The digital signatures are added by a trusted third party, called the certificate authority (CA). Certificate author-ities confirm the relationship between individuals and their public keys. The X.509v3 certificates are used as the globally unique principal creden-tials. Some of the standard fields of X.509v3 (like the organization field in the distinguished name) are used as security attributes. X.509v3 certif-icates can also include extensions that provide a mechanism to store attributes that are not represented in the standard fields.[12]

The Java Cryptographic Extensions package is used to apply and check the cryptographic protections.

These certificates are stored in DNS CERT records. This provides a secure distributed storage-and-retrieval mechanism for the certificates. Each certificate can be located by the domain name of its CERT record. Henceforth, the fully qualified domain name of the issuer's certificate was included in each certificate, to aid in distributed certificate tree processing. As a result, when an active packet travels through the network, it is always possible to trace the certificates needed to fully verify the certificate signatures.

The keynote policy system allows end users to include their own authorization policies (e.g., for state created in the infrastructure) in the active packets. These policies are enforced through the security architec-ture of Java 2. Active code was loaded with a class loader. The class loader created the protection domain for the active code with a permission object containing the credentials associated with the active packet. The original ANTS packet contains a header and the data payload. Some fields of this header are variable for each node, such as a time-to-live (TTL) field. Some fields remain static throughout the packet's travels, such as the MD5 hash identifier of the active code. The Active Network Encapsulation Protocol (ANEP) packet format is shown in Figure 9.3.

Hence, the packet payload has been divided into two sections; one is a static area (covered by the authentication protection) and the other is a variable area. The static area of the packet consists of the static portions of the EE header, for instance, the code identifier, and the static portions of the data payload. The variable area of the packet includes the variable fields of the EE header and the variable portions of the data payload. The active code itself cannot be modified while in transition in the network.

ANEP Header	Credential Field	Static Payload	Origination Signature	Varying Payload	Hop Integrity	ANEP Options

Figure 9.3 The ANEP packet format.

When a packet arrives, the credential references are extracted from the packet. The associated X.509v3 certificates are retrieved from DNS CERT and their signatures verified. This requires the recursive retrieval of certificates associated with the issuer of the certificate, just like a DNS for the Web. When the certificates have been checked, the packet signatures are verified at the end of the process. If this is successful, the packet is authenticated.

The authentication handling should be performed in the NodeOS so that access to cut-through channels can be authorized according to the correct domain, and the EE is not inundated with unauthorized packets. If the EE senses that it is under attack, which can create congestion, it may restrict that policy. So there must be NodeOS API calls to modify a domain's access control policy.

9.4.3.2 Authorization Process

The authorization developed for SANTS is based on the Java 2 security architecture, with certain modifications to support the special requirements of the active network environment.

Each class in Java 2 is provided a protection domain by its class loader. When a SANTS class loader is called upon to load active code, it binds the active code class to a SANTS protection domain object that contains exactly one SANTS permission object. This SANTS permission object is instantiated and initialized with a credential set belonging to the principals associated with the SANTS class loader. In turn, whenever the SANTS permission object is called upon by the access controller to render an access decision, it consults the SANTS policy engine, which determines whether the requested resource access is allowed by the access policy.[4] If any of the key owners (users) are authorized according to the policy, the access succeeds.

9.4.4 Secure QoS Handling (SQoSH)

SQoSH[5] extends SANE to support restricted control of quality of service in a programmable network element. It has a central element called the Piglet. This is a lightweight device kernel that provides a virtual clock type of scheduling discipline for network traffic, and the clock is adjustable. It controls access to managed resources and integrates this control with the resource management mechanisms provided by the Piglet operating system. Figure 9.4 shows the SQoSH architecture.

SANE is the upper layer in this architecture, which provides access to resource management of Piglet. SANE also performs security checks

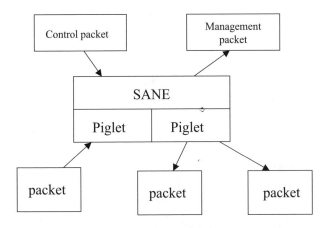

Figure 9.4   The SQoSH architecture.[5]

through cryptographic operations required for granting access to the Piglet resources. The Piglet then decides whether the resources can be allocated to the validated request. Packets destined for SANE are demultiplexed by the Piglet, which provides basic packet communications for SANE. A complete SQoSH system would consist of a multiprocessor with a Piglet instance on each device-managing processor.

The Piglet architecture runs the *lightweight device kernel* (LDK). In this system, one or more system CPUs are dedicated to continually run the LDK. The LDK implements only the minimal set of functions necessary to support direct user-level access to physical devices, namely, protection, translation, and multiplexing. This technique reduces the kernel response time because it communicates with devices through polling rather than interrupts. Furthermore, applications activate kernel services using shared memory communication, which incurs much lower overhead than the conventional system call trap.

The SQoSH architecture facilitates safe and secure access to network resources, allowing end users to manage these resources. It also offers controlled access of system resources by using the Piglet operating system, eliminating a large class of denial-of-service attacks.

9.4.5 PLAN

Programming Language for Active Networks (PLAN)[6] is an example of a language particularly deployed for active networks that deals with the security and safety issues from a programming viewpoint. These programs substitute the packet headers used in present networks. The basic design consideration for PLAN is to have programs that are lightweight and of controlled functionality. In spite of this controlled functionality, the limitations on the capabilities of PLAN are somewhat relaxed by allowing PLAN code to call other service routines that reside in nodes and are written in other, more powerful languages. However, an authentication procedure is adopted for these service routines. PLAN addresses the issues of safety and security, performance, and flexibility as follows:

- **Safety and security**: The requirements on bandwidth, processing time, and memory are addressed in two ways. First, PLAN programs are guaranteed to terminate. This is so because recursive function calls and unbounded iterations are absent from the language. Second, PLAN programs have bounds on the amount of resources that they can consume. Third, PLAN is a purely functional and strongly typed language. Thus, PLAN programs are statically type checked before they are injected into the network, so that they do not have type errors. Additionally, PLAN programs are pointer safe and concurrently executing programs cannot interfere with each other.
- **Performance**: A major benefit of keeping PLAN simple is that its execution is lightweight and common tasks can be done easily and quickly.
- **Flexibility**: PLAN is not completely general but is able to express programs for network configuration and diagnostics, and to provide the distributed computing middleware type of functionality that connects router resident service routines across different nodes and protocols.

PLAN was used in the PLANet.[6] PLANet is an active internetwork in which all packets are PLAN programs. Service routines are also supported.

9.4.6 Active Edge Tagging (ACT): An Intruder Identification and Isolation Scheme in Active Networks

9.4.6.1 Background

Distributed denial-of-service (DDoS)[7,13] attacks typically occur when a large number of hosts simultaneously attack a site. This type of attack has been known for some time, but defending against it has been a challenging problem. One of the more problematic issues concerning these attacks is isolating the attack packets from reaching the victim's sites or identifying the intruders' locations, because forged (or spoofed) IP source addresses are used.

There are a number of solutions presented over a period to help alleviate problems caused by DDoS attacks. However, these solutions have proved to be limited, and as a result, source address-spoofing denial-of-service attacks are still a major threat in the Internet.

One of the ways to counter these types of attacks is ingress filtering,[14] which discards packets that do not match specific conditions set forth in the access lists of access routers. When properly configured and ubiquitously supported by network operators, ingress filtering can effectively prevent this type of attack. However, it has not been very popular among the community of network operators due to its costly operations and the effects of misconfigured access lists. In addition, the effectiveness of ingress filtering greatly depends on the relative location of network elements (NEs) or network nodes implementing it.[7]

There are other methods proposed to help mitigate source-spoofed denial-of-service attacks, such as egress filtering and Internet Control Messaging Protocol (ICMP) traceback messages.[14] However, all of these schemes have not been deployed by most network operators due to their restrictions and requirements. As a result, source address-spoofing denial-of-service attacks remain one of the main threats.

G. Kim et al.[7] have presented a novel solution called active edge tagging (ACT) for DDoS attacks. ACT facilitates handling of source-spoofed attacks by effectively detecting, identifying, and isolating intrusions at the network layer. In contrast to the existing solutions, ACT deals with the targeted attacks efficiently without compulsory participation of every individual network in the Internet. ACT is particularly effective for identifying and isolating attackers employing a DDoS type of intrusion scheme, and it is highly scalable and extensible. Hence, it is feasible to implement in large-scale networks such as the global Internet.

9.4.6.2 Details of ACT

ACT is an intruder identification and isolation scheme for active network environments. First, it identifies the physical location of attackers that disguise their identities with spoofed IP source addresses. Second, it isolates the attackers from potential victims by discarding the ill-intended packets before they reach their target. ACT is primarily intended for DDoS types of intrusions, where multiple packets using spoofed IP source addresses are transmitted over an extended period. ACT can also be employed for non-DDoS attacks that use forged source addresses. Active networks provide an effective environment for edge tagging by enabling customized installation of programs and packet manipulations in network elements.[7]

The source address field in the IP packet's header is used for identification of the source of the packet. The inclusion of source and destination addresses in each packet is necessary for proper routing, as well as interactions between a client and a server in a distributed operation.[7] When an intruder hides attackers' identities with randomly forged source addresses, the attacked node has no means of locating the attack agents without extensive investigative searches and analysis based on the transaction logs, which may take a long time. Additionally, there is no guarantee of identifying the intruder in this case.

ACT identifies the physical location of attackers' agents when spoofed source addresses are used. The objective of ACT is to take advantage of the first-hop routers' capability to identify the actual 32-bit IP source address of each incoming packet from host machines. This approach simplifies the network operations greatly by not requiring mandatory participation from every individual network for filtering or monitoring attack packets. An attacked node can actively initiate the attack handling procedure by activating ACT. In other words, in contrast to the static nature of other attack-fighting schemes, ACT can be dynamically invoked only when the early signs of attack start to appear, eliminating the need for continuous monitoring of traffic.

In ACT, an ingress router tags the actual 32-bit IP source address in the active network layer header (ANEP level) for the packets destined to the victim. A monitor at the egress point of the network examines the packets destined to the attacked destination. For each packet found, the monitor checks whether the tagged address (actual IP source address) carried in the active network layer is consistent with the IP source address carried in the packet's IP header. If the two are different, the packet is regarded as a malicious packet, and thus it is classified as an attack. Once an attack packet is detected, its actual IP source address is logged for future investigation, and the packet may be discarded before it reaches the intended attacked destination.

To minimize the communications and resource overhead in ACT, edge tagging is activated on a per domain basis until an ample number of attack agents' locations are found. In addition, edge routers are activated over multiple phases, each separated by a predefined period, and only a partial set of edge routers is activated in each phase in each domain. G. Kim et al.[7] have considered a simple environment with a single-domain network consisting of a set of active routers. They have also assumed that the edge routers correspond to first-hop routers from the end user's point of view.

ACT ensures scalability such that ACT works seamlessly and efficiently in large-scale multidomain networks, including wide area network (WAN) or global Internet, where a large number of different network domains are interconnected, each of which potentially consists of many edge routers. The summary of key features of ACT is as follows.[7]

1. The primary goal of ACT is to identify attackers' physical locations, keeping in view the scalability and other overhead concerns. The secondary goal is to isolate the attackers by discarding attack packets before they reach the destination victim premises.
2. Edge tagging is performed on a per domain, per need basis. Edge routers in each domain are activated over multiple phases, each separated by a predefined period, and only a subset of edge routers are activated in each phase. Per-domain-based activation and traffic monitoring help to achieve scalability in terms of the size of the protocol database, traffic amount, and the number of edge nodes handled by a single monitor.[7]
3. The start of edge tagging is explicitly signaled by a local monitor. An internal timer, however, triggers the completion of edge tagging. This implicit completion of edge tagging eliminates protocol message overhead for signaling the end of edge tagging. The completion of traffic monitoring in each domain is also triggered by an internal timer.[7]
4. A few performance-critical parameter values can be adjusted, depending on the size of the network deployed, the average size of a domain, and the type of attacks primarily targeted.

    ■ **Depth of recursion per activation cycle**: This parameter identifies the maximum number of attackers associated with the number of edge routers activated. The depth size depends on the scale of the network and the type of attacks. The maximum level of recursive activation depth at each activation phase can be calculated. The depth of recursion is related to the total number of activated edge routers. With other parameters fixed, an increase in the depth of recursion in a single activation cycle increases the total number of activated edge routers within a single cycle of activation. Because the goal is to identify the maximum number of attackers with the minimum number of edge routers involved, the proper choice of depth depends on the scale of the network and the type of attacks.

    ■ **The number of edge routers activated per activation cycle**: A local monitor in a network activates its edge routers over multiple activation cycles. The reason behind this is to find an intruder with a minimum number of edge routers involved in edge tagging. Thus, the response time depends on the number of edge routers activated in each activation cycle; this parameter can be adjusted according to the network environment and the type of attacks.

■ **Duration of edge tagging/monitoring**: The completion of edge tagging and traffic monitoring is initiated by local timers. Timer values are important for system performance because different types of attacks can use different time durations to repeat generating packets in an attack. Thus, the duration of edge tagging and packet monitoring should be long enough to cover multiple cycles of packet generation. A set of different timer values can be chosen for different types of attacks in this scheme.

ACT, with the help of an active network environment, does not require mandatory participation of every individual network in the Internet. This reduces the load and complexity in networks.

9.4.7 Active Security Support for Active Networks

This security architecture was presented by Liu and Campbell.[8,15] The secure node architecture in this scheme is integrated into the operating system of an active node for the purpose of securing active network infrastructure. It supports dynamic security services and highly customized policies created by users and active applications. The architecture consists of the following components:[8]

■ An active node operating system (NodeOS) security API
■ An active security guardian
■ Quality of protection (QoP) provisions

As shown in Figure 9.5, the key components are the security API, the security guardian, and the QoP provisions.

The security API supports authentication, authorization, integrity, and access control to EEs and active applications. To support flexible, distributed, and dynamic security policies, applications, or policy, network administrators use a customized code fragment that encodes the type of access control policy and other constraints used in the access control decision-making process. This code fragment is called the *active capability* (AC).[8] Essentially, an AC is executable code that concisely represents dynamic security policies and mechanisms. The security guardian evaluates ACs in a secure sandbox environment and enforces the security requirements of AC evaluation results. It obtains ACs securely through the support of the AC communication protocol. To support application-specific security and

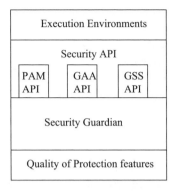

Figure 9.5   The key components of secure node architecture.

protection requirements, the secure node architecture provides quality of protection provisions for customized security services.

9.4.7.1   The NodeOS Security API

The NodeOS security API makes security services such as authentication, authorization, and integrity available to EEs and active applications. The API needs to be as generic as possible to accommodate a wide variety of applications.

The NodeOS security API has three major parts: the authentication API, the authorization API, and the security services API, as shown in Figure 9.5.

- The authentication API authenticates execution environments (EEs), active applications (AAs), or users. It is based on the pluggable authentication module API (PAM API).[8] This module uses the X.509 public key infrastructure (PKIX).
- The authorization API provides protection. It is based on the generic authorization and access control API (GAA API).[8] The security guardian in Figure 9.5 supports access control policy evaluation and enforcement. The security guardian's functionality is the same as that of a traditional reference monitor, or like that of a supervisor privilege in a traditional operating system. All accesses to node resources must go through the security guardian. One possible implementation of an access control mechanism is the active capability described later.
- The security services API offers encryption, digital signatures, and protection requirements. The security services API is based on the generic security service API (GSS API)[8] and was extended to support quality of protection, discussed later. The GSS API provides

generic security services to application programmers. It performs two operations. First, it creates a security context between communication peers through peer entity authentication. The second operation is on a per message basis. It may include confidentiality, integrity, and data origin authentication, depending on the application-specific security requirements.

9.4.7.1.1 Quality of Protection

By using NodeOS security API, the security guardian, and active networking features like execution environments, it is possible to support quality of protection provisions for applications. QoP provisions support customized, flexible security and protection requirements of applications.

To provide quality of protection, the NodeOS API needs to be enhanced with different security and protection options. These options are supported by the underlying security library implementation in the NodeOS. In addition, they are based on security and protection features with certain characteristics. The QoP has the following properties:[8]

■ Key length of security algorithms.
■ Robustness or strength of security algorithms.
■ Security mechanisms for authentication and privacy.
■ Trust values for developers/vendors of security implementations.
■ Assurance level of a router NodeOS: a router NodeOS with higher assurance class is trustworthier.

Active capabilities specify, control, and manage QoP. A trusted party creates ACs when requested by applications.

With a NodeOS security API, a security guardian, and QoP provisions, the secure node can provide active security features to applications. QoP provisions permit trade-offs between security protection and satisfaction of the QoS constraints through dynamic reconfigurations. The protection may be provided on a per flow, per capsule, or per service basis to optimize performance overhead based on application needs. Through these methods, a security-customized routing path can be specified by an application; this path will be protected through stronger protection under intrusion.

9.4.7.2 Active Access Control

Active access control facilitates the creation of a customized piece of code that instructs the type of access control policy. This code fragment is called the *active capability* (AC).[8] A security guardian is implemented in the NodeOS to evaluate and enforce ACs. Applications can use ACs to encapsulate credentials and encode situational policies that are authenticated by a trusted policy server to add, alter, or revoke existing access control rules and mechanisms dynamically. Applications write active network code that uses these credentials and policies to inject customized security into the routers. Thus, applications may choose an access control policy and enforce this policy on their active network code. The implementation can provide consistent security policy guarantees and platform-independent enforcement of security policies across all active routers spanning the network.

Active capabilities[18] are used to support flexible distributed dynamic security policies. Conceptually, an active capability is a piece of unchangeable code that encodes a critical, application-specific part of the decision-making code used in access control. An active capability is an executable Java bytecode that concisely represents dynamic security policies and mechanisms. In addition, an active capability is protected by digital signatures and can reside in user space and is freely communicated around.

9.4.7.2.1 Policy Framework

This framework supports ACs and enables users to have flexibility in terms of policy specification. This is implemented through an object-oriented policy representation framework in Java. It allows users and commercial organizations to specify policies tailored to their specific operational needs. The framework[8] itself is a hierarchy of classes, as shown in Figure 9.6.

The framework is dynamically configurable and extensible. The classes at the lowest level of the framework are mostly abstract and are mainly used to represent concepts such as mappings. These classes establish the basis for a hierarchy of successively more and more specialized classes, representing concepts such as labels and access control lists. Lastly, at the top of the framework are classes that offer a variety of generic policy forms.

9.4.7.2.2 The Security Guardian

The aim of the security guardian in the architecture is to support AC evaluation and enforcement. All accesses to node resources must pass through the security guardian. The signature of the active capability is checked by the security guardian.

The security guardian's functionality is similar to that of a traditional reference monitor, but with several major differences. In traditional systems, a reference monitor is interposed between the subjects and objects to control subjects' access to objects based on access authorizations. The

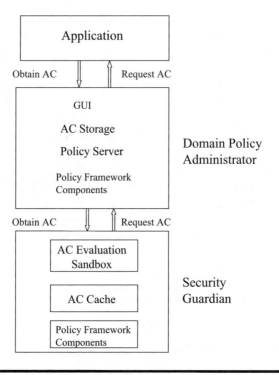

Figure 9.6 Policy administration framework.[8]

traditional reference monitor is passive. There is no need to execute and evaluate an operation request message. The traditional passive reference monitor is simply an enforcement engine, and its power and functionality are limited.[19]

The security guardian enforces the security checks by specifying the preconditions and postconditions. Preconditions permit necessary security checks to be performed before the actions take place, and postconditions can be used to maintain state and perform additional checks after the action has been completed and when more information becomes available. An administrator graphical user interface (GUI) is present as a front end to the policy framework.

9.4.7.2.3 Policy Administration

Policy administration consists of policy servers that operate as a communication front end for distributing executable security policies in the form of ACs. An AC may either provide all the code for a security policy or access control or specify a policy server from which to retrieve code. The security guardian provides the necessary security access controls to resource requests from users. As the access controls are applied, the AC may request further policies that must be downloaded from a policy server.

The policy administrator and policy server are trusted entities. In this architecture, they perform the role of a trusted third party. The security guardians are also trusted entities and execute in trusted kernel space. The active capabilities are used to distribute policy descriptions. The complex policies and controls from the framework can be securely downloaded on a component basis from the repository into the security guardians. This is achieved by using a secure communication channel. The framework uses the policy server to accept AC requests from either the security guardian or applications, and to provide the requested customized AC. Typically, one or more policy servers are associated with each protection domain. Application programs or security guardians contact their closest or least-loaded server and obtain the active capability on the fly. If the necessary AC is not put into the active packet by applications, the security guardian of the first router of a protection domain needs to obtain the AC from a policy server.[8] For performance and load reduction reasons, the security guardian can put the obtained AC into the active packet so that there is no need to request the AC dynamically at each router of the domain.

To improve the AC evaluation efficiency, the security guardian uses a cache to store the ACs, or even the results of AC evaluations. Depending on the freshness and type of AC, a request may be satisfied by a simple cache lookup instead of an expensive AC evaluation. On the other hand, for some types of capabilities, the security guardian can always download the latest capability from the policy server. Caches are flushed out periodically to maintain their freshness. These caches improve the performance, but they lead to the problem of cache consistency.[8]

Another important attribute of this architecture is the ability of the trusted authority to cancel a capability at any point in time. The trusted authority can flush out the cache by sending a message to the relevant security guardians and install a new capability at runtime. Alternately, the application can present a properly signed new capability on the fly with a newer version number, which cancels the existing capability.[8]

9.4.7.2.4 Low-Level Code Safety

The evaluation engine of the security guardian depends on the Java bytecode verifier[8] for low-level code safety. Before loading a class, the verifier performs data flow analysis on the class code to verify that it is type safe and that all control flow instructions jump to valid locations.[8]

There are many other approaches for low-level code safety. The PLAN project[6] uses programming language techniques to address the code safety problem. Capsules are written using a strongly typed, resource-limited language, and dynamic code extensions are secured by using type safety and other mechanisms. Another approach is proof-carrying code (PCC).[8] Besides regular program code, PCC carries a proof that the program satisfies certain properties. The proof is verified before the execution of the code. The generation of a proof may be complex and time-consuming, although its verification should be simple and efficient. Software fault isolation (SFI)[8] provides another alternative for low-level code safety. It uses special code transformations and bit masks to ensure that memory operations and jumps access only the right memory ranges.[8]

In short, there are a variety of different mechanisms and protocols proposed. Each method has its own benefits and drawbacks. Ultimately, the application must be given the choice to pick the mechanism that is most appropriate for its use. The secure node architecture should be general enough to allow all these mechanisms to coexist.

9.4.8 The Dynamic Access Control Mechanism

This scheme has been presented by A. Hess and G. Schäfer.[9] It makes sure that only authenticated code can be executed on an active node. A service can be signed only by the owner of the service or the network administrator. The presence of the administrator's signature indicates that the service security policy has been inspected by the administrator and a trust label for the service is specified by the owner. The trust label is a means to specify the credibility of a service, while it simultaneously determines the minimal access control mechanism configuration for this service (at least the set of access control kernel modules specified by the corresponding trust label must be loaded before the execution of the service can be started). Figure 9.7 shows the access control framework. It consists of:[9]

- A security daemon
- A policy handler
- A set of access control kernel modules (ACKMs)
- A node security policy
- A service revocation list (SRL)
- A list of default authorization and resource descriptions (DRAD)

In the list of components given above, the security daemon and policy handler are the only permanent entities. The remaining entities are activated on demand.

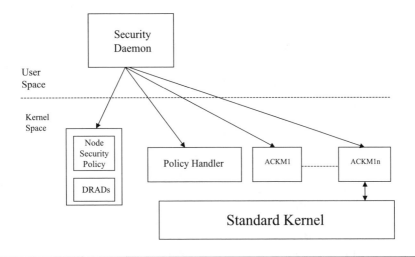

Figure 9.7 The access control framework.

9.4.8.1 The Security Daemon

The security daemon performs the following tasks:

- Verification of digital signature
- Comparison between service and node security policy
- Activation/deactivation of ACKMs
- Handover of service security policy to the policy handler module

9.4.8.2 The Policy Handler

The policy handler takes care of the following procedures:[9]

- Supervision of the activation/deactivation of ACKMs
- Supervision of the exit phase of services
- Management of the service security database

The policy handler monitors the exit phase of each service running on an active node. If a service exits, it must be guaranteed that all child processes also terminate. Hence, the policy handler keeps track of all child processes of each service.[9] In case a child process does not terminate and the active service has already ended, the policy handler itself terminates the child process.

The policy handler performs the management of the service security database. A user space process is identified through its process identification number (PID). The PID and service security policy are linked and afterwards inserted into the database. If a service exits, it is the policy handler who removes the corresponding entry from the database.

9.4.8.3 *An Access Control Kernel Module (ACKM)*

Each ACKM is accountable for the supervision of one specific resource. The supervision is based on the technique of intercepting system calls. By incorporating a loadable kernel module, it is possible to stretch the functionality of the kernel. A process must execute a system call to get access to the operating system services. An ACKM intercepts exactly one system call. An ACKM checks the database for the service security policy related to the requesting process, to verify if the process is within its resource limits or if the service is authorized to use a specific operating system service. If this test is passed, the ACKM updates the appropriate data structure in the service security database to keep track of the resource utilization of each service.[9]

9.4.8.3.1 Node Security Policy

The node security policy defines which resources and operating system (OS) services are generally accessible by the services on each active node. Furthermore, it defines the complete amount of each resource that can be maximally consumed in aggregate by all actual running services. The node policy provides a means to bound the set of services that can be executed on an active node.

A service listed in the service revocation list (SRL) must not be executed on any active node, and each default authorization and resource descriptions (DRAD) entry specifies the authorization and resource limits for services of a specific trust label.[9]

9.5 Summary

Network security is at the heart of any communication system. It becomes more critical in case of active and programmable networks, because they open up nodes (routers, etc.) to allow dynamic downloading of code through packets or agents. Different research projects dealing with the security of active and programmable networks have been discussed. Some of the major security threats and concerns have been identified as well.

SANE consists of active packets that carry mobile active codes. They change the functionality of network elements and can be dynamically loaded. SANE also consists of a secure bootstrap architecture that provides a high-integrity base on which the security of the other layers depends. The SANE architecture performs dynamic and static checks.

Safetynet deals with the issue of designing a programming language for active networks. It presents a programming paradigm that protects the connectivity of the network and places specific requirements to ensure security, safety, and fair resource allocation.

SANTS, based on ANTS, created an execution environment supporting strong security by extending the ANTS execution environment. It uses X.509v3 certificates for authentication. When a packet arrives, the credential references are extracted from the packet. The associated X.509v3 certificates are retrieved from DNS CERT and their signatures verified.

SQoSH extends SANE to support restricted control of quality of service in a programmable network element. It implements an operating system called Piglet. It controls access to managed resources and integrates this control with the resource management mechanisms provided by the Piglet operating system. This saves the system from denial-of-service attacks.

Programming Language for Active Networks (PLAN) is an example of an active network language that deals with security and safety issues from a programming viewpoint. These lightweight programs substitute the packet headers used in traditional networks. PLAN is a purely functional and strongly typed language. Hence, PLAN programs are statically type checked before injection into the network

The chapter also describes the active edge-tagging (ACT) intruder identification and isolation scheme in an active network environment. ACT identifies the physical location of attackers who hide their identities with spoofed IP source addresses.

The active security architecture by Z. Liu et al.[8] supports dynamic security services and highly customized policies created by users and active applications. It consists of an active node operating system (NodeOS) security API, an active security guardian, and quality of protection (QoP) provisions.

The dynamic access control mechanism presented by A. Hess and G. Schafer[9] makes sure that only authenticated code can be executed on an active node, and that a service can be signed only by the owner of the service or the network administrator. The presence of the administrator's signature shows that the service security policy has been inspected by the administrator and that a trust label for the service is specified by the owner.

Exercises

1. What are the major security threats in networks? In what respect are they different from the threats in active and programmable networks?
2. Discuss the security architecture of SANE. Explain the function of the AEGIS layered boot and recovery process.
3. What type of support does Piglet provide to upper layers in SQoSH?
4. Compare the properties of Safetynet and PLAN.
5. Discuss the authentication process in SANTS.
6. Explain how active edge tagging (ACT) deals with distributed denial-of-service attacks in active networks.
7. What is the function of the security daemon and policy handler in the dynamic access control mechanism?

References

1. D.S. Alexander, W.A. Arbaugh, M.W. Hicks, P. Kakkar, A.D. Keromytis, J.T. Moore, C.A. Gunter, S.M. Nettles, and J.M. Smith, The SwitchWare active network architecture, *IEEE Network*, 12:3, 29–36, 1998.
2. D.S. Alexander, W.A. Arbaugh, A.D. Keromytis, and J.M. Smith, A secure active network environment architecture: realization in SwitchWare, *IEEE Network*, 12:3, 37–45, 1998.
3. K. Psounis, Active networks: applications, security, safety, and architectures, *IEEE Commun. Surv.*, 2:1, 2–16, 1999.
4. S. Murphy, E. Lewis, R. Puga, R. Watson, and R. Yee, Strong Security for Active Networks, paper presented at IEEE Open Architectures and Network Programming (OPENARCH), Anchorage, AK, USA, pp. 63–70, 2001.
5. D.S. Alexander, W.A. Arbaugh, A.D. Keromytis, S. Muir, and J.M. Smith, Secure quality of service handling: SQoSH, *IEEE Commun. Mag.*, 106–112, 38:4, 2000.
6. M. Hicks, P. Kakkar, J.T. Moore, C.A. Gunter, and S. Nettles, PLAN: A Packet Language for Active Networks, paper presented at the Proceedings of the International Conference on Functional Programming (ICFP), Baltimore, MD, USA, pp. 86–93, 1998.
7. G. Kim, T. Bogovic, and D. Chee, ACtive edge-Tagging (ACT): An Intruder Identification and Isolation Scheme in Active Networks, paper presented at the Proceedings of the Sixth IEEE Symposium on Computers and Communications (ISCC'01), 2001.
8. Z. Liu and R.H. Campbell, Active security support for active networks, *IEEE Trans. Syst. Man Cybernetics C Appl. Rev.*, 33:4, 432–445, 2003.
9. A. Hess and G. Schäfer, *Realization of a Flexible Access Control Mechanism for Active Nodes Based on Active Networking Technology*, IEEE Communication Society, 2004.

10. D.L. Tennenhouse, J.M. Smith, W.D. Sincoskie, D.J. Wetherall, and G.J. Minden, A survey of active network research, *IEEE Commun. Mag.*, 35, 80–86, 1997.

11. ITU Recommendation X.509, ISO/IEC 9594-8: 1995, Information Technology — Open Systems Interconnection — The Directory: Authentication Frame, 1997.

12. E. Gerck, Overview of Certification Systems: X.509, CA, PGP and SKIP, available at http://www.mcg.org.br/cert.htm.

13. S. Karnouskos, Dealing with Denial-of-Service Attacks in Agent-Enabled Active and Programmable Infrastructures, paper presented at the Proceedings of the 25th Annual International Computer Software and Applications Conference (COMPSAC.01), 2002.

14. P. Ferguson and D. Senie, Network Ingress Filtering: Defeating Denial of Service Attacks Which Employ IP Source Address Spoofing, IETF RFC 2827, May 2000.

15. Z. Liu, R.H. Campbell, and M.D. Mickunas, Securing the node of an active network, in *Active Middleware Services*, S. Hariri, C. Lee, and C. Raghavendra, Eds., Kluwer, Boston, 2000.

16. R.S. Sandhu and E.J. Coyne, Role-based access control models, *IEEE Comput.*, 29:2, 38–47, 1996.

17. The SwitchWare Project Homepage at the University of Pennsylvania, available at http://www.cis.upenn.edu/~switchware/.

18. Z. Liu, P. Naldurg, S. Yi, T. Qian, R.H. Campbell, and M.D. Mickunas, An agent based architecture for supporting application level security, in *Proc. DARPA Information Survivability Conf. Expo. (DISCEX'00)*, Vol. 1, Hilton Head Island, SC, January 25–27, 2000, pp. 187–198.

19. M.D. Abrams and J.D. Moffett, A higher level of computer security through active policies, *Comput. Security*, 14, 147–157, 1995.

Chapter 10

Applications of Active and Programmable Networks

10.1 Introduction

The Internet has already grown into a meganet with global reach.[1] With the emergence of advanced applications, it is now poised to evolve into a complex system of systems. With its expansion, the asymmetry of the Internet is also increasing. Historically, the initial Internet architecture was conceived to cope with the heterogeneity of network standards.[2,3] No sooner had we thought that the problem was solved than a second era started evolving. The next generation of the Internet will have to deal with more intrinsic heterogeneity — the asymmetry of hard network resources such as bandwidth or switching capacity. Network technology that facilitates domain-specific processing within a network may provide novel advantages for the adaptive applications of this generation. The commercial importance of the adaptive system is growing in several areas, particularly in scalable video communication, Web caching, and content adaptation services. The advent of mobile information systems has generated another wave. Due to the lack of network support, the first-generation techniques are compelled to depend on resilience (by means of redundancy or indirect application-level network impairment probing) to survive variation in transport characteristics and edge processing.[4,5]

This chapter discusses several applications within the context of active and programmable networks. Examples are active e-mail,[6] distributing of video over the Internet using programmable networks,[7–9] and active traffic and congestion control mechanisms.[10–12]

10.2 Active Electronic Mail

One of the most widely used computer communication tools today is electronic mail (e-mail). E-mail has been on the scene since the earlier days of computer networks and has become one of the three "killer applications," along with Telnet and File Transfer Protocol (FTP).[6] Its popularity and ubiquity have established it as a communication standard worldwide. The underlying network infrastructure has more to offer, as it is more intelligent and service oriented. However, e-mail does not generally take advantage of the new advances in these domains.

10.2.1 Active E-Mail Infrastructure

Enhancements on the current e-mail platform can be done on the server side, on the client side, and on the network components that exist in the server and the client side.

10.2.2 User Context Awareness

Traditionally, e-mail is delivered to the user's e-mail server. The mail client pulls the server, gets the e-mail, and then, depending on its capabilities, tries to present it to the user via a single device. However, in the future e-mails will be pushed to the user according to his or her profile and context. For example, suppose that you receive a multimedia e-mail (with video and sound), but unfortunately your end device is a personal digital assistant (PDA) with no sound support. The network recognizes dynamically from your context that you also have a mobile phone. Therefore, it sends the video to your PDA and the sound to your mobile phone.[6] The user's context includes not only location, but also capabilities of devices around the user that can be controlled. Therefore, the network should be aware of the user's context, i.e., it dynamically discovers the user's position, preferences, devices, and obviously the capabilities of this specific context at that specific moment. The management and acquisition of this kind of information can be assigned to the agents. By querying devices such as active badges, Global Systems for Mobile (GSM) phones, etc., the agents are in a position of estimating the user's position and maintaining the dynamic changes to the user's profile database. Based on the profile stored in this database, the agents residing on various nodes within the network are able to adapt their behavior while processing the e-mails directed to that specific user. The dynamic update of the user's context is difficult; therefore, technology has to take as much responsibility as possible from the user and delegate it to intelligent agents, which will behave autonomously.

10.2.3 Distributed Antispam

Spam e-mails are generally defined as unsolicited messages to many people. This type of unwanted communication floods the Internet. Beyond reducing the bandwidth by filling up communication links, they also create annoyances to many companies by hindering corporate users from their daily tasks when they try to read and delete them. There are some tools that allow filtering of e-mails on the client side, and also some server-based ones; however, these tools are not personalized and are based mostly on keyword and IP address filtering.[6] Thus, a solution is there, but it is awkward, nonintelligent, and too complicated for the everyday user who just ignores it.

As a result, frustration is on a daily basis, and this has several economic and social side effects for companies. The problem is that users have little control over the e-mail they receive, and agent-based active networks are able to help them in this matter. The e-mails that pass through the node can be filtered and ones corresponding to well-known spammers and related Web sites can be dropped. Real-time checking can be done today, as there are sites that maintain blacklists of open mail relays, dial-up lists, and Web sites. Filtering of e-mails can be performed based on user preferences, as they are defined in a profile database. The user may for personal reasons decide to deny e-mails coming from specific domains or people.

Similar to that, e-mails recognized and classified as spare trigger the appropriate actions, e.g., delete, notification of the spammer's mail provider, etc. The combination of the two methods would be ideal, as it introduces customized blacklists based on user preferences. These anti-spare tactics can be invoked in several places in the active e-mail infra-structure. They can be invoked at the sender's message transfer agent (MTA), but usually the spammers set up their own MTAs, and large Internet service providers (ISPs) prefer loose enforcement of antispam techniques for fear of accidentally losing e-mails due to false alarms. Therefore, the best place would be within the network that is in the MTA relays.[6]

10.2.4 Mail Storage

The received e-mail is usually stored as text. This makes tasks such as search, categorization, etc., very difficult and complicated. Almost everyone keeps e-mails in the incoming mailbox for future reference or as reminders of tasks they have to complete. Therefore, it would be useful if the e-mails of the user could be used as a knowledge base. E-mails should be stored in a universally accepted format so that applications can have easier and better results while processing them. One way would be to store them as files in a database where structured queries can be done. The eXtensible Markup Language (XML) comes with a variety of tools and is driven by the vision of a universal homogeneous view of information.

10.2.5 Mail Notifications

Currently e-mail is passive. You have to check your e-mail (pull method) and see whether you have new messages. However, activeness is required. The e-mail infrastructure should notify you when something happens (push method) based on your profile. The agents can intelligently monitor your mailbox and, based on your preferences, send notifications or the e-mail itself when it is needed. In this way, the interaction with the user is dynamic and may spawn agent-wrapped services, e.g., Instant Messaging[6] or even automatic replies. Furthermore, this approach scales well for integration of multiple-user mailboxes in various ISPs without user intervention.

10.2.6 Mobility

With the emergence of mobile and palmtop computing, support of nomads via intelligent muting and downloading of information such as e-mails to pagers, cellular phones, and palmtop computers is mandatory. People want to be offered seamless access to their e-mails from any location. Today's e-mail infrastructure does not take care of mobility and other requirements, e.g., location awareness, bandwidth, and device dependencies and capabilities.

By using active and programmable nodes, the route of information for a user-specific flow at various levels can be changed. If an active network node does support queries to the user profile, it can consult it to redirect the e-mail to the current location of the user, convert it to another media form, e.g., fax or SMS (short message), or even drop it because this e-mail's usefulness depends on a user-specific geographic location.

10.3 Distribution of Video over the Internet Using Programmable Networks

The significant bandwidth requirement of video makes large-scale video broadcasting with QoS difficult in today's Internet. Because bandwidth guarantees are not possible, the transmission process must adapt in real-time to the available bandwidth. Such adaptation must take into account the heterogeneous nature of client connections to the Internet, to ensure that each client receives the video at an acceptable level of quality.

The self-organizing video streaming[8] scheme addresses adaptation of Motion Picture Experts Group (MPEG) 2 with respect to two critical network resources: (1) bandwidth at various links and (2) the processing resource at the junction nodes. The key to adaptive streaming is a mechanism for active transcoding. For link capacity adaptation, the mechanism senses local asymmetry in link capacities at various junction points of a network. Accordingly, it adapts the video stream rate. For the second resource, the mechanism senses the local computation power. Based on the network computational power, it demonstrates self-organization behavior. Figure 10.1 shows the setup for self-organizing video streaming.

This scheme uses a combination of legacy and active routers. In this environment the passing video stream appears as a self-organizing stream that automatically senses the network asymmetry and adjusts as the packets travel via the active subnet. The thinly distributed active nodes provide

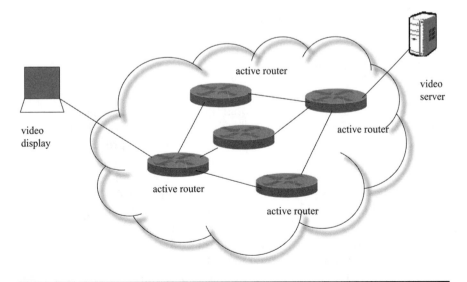

Figure 10.1   The setup for self-organizing video streaming.

the computing platform required for the conversion. The system is adapted at two levels:

1. Modular self-organization
2. Computation diffusion

These two techniques represent different levels of adaptivity within the network. To adapt to the available computation resource, the first uses a domain-specific technique to cut back on the internal computation of a node. When the capacity of a single node is too small to support the transformation, the second approach tries to diffuse the computation onto the multiple nodes in the network neighborhood.

In another scheme for video transmission over active routers,[7] the network architecture consists of both active and legacy (nonextensible) routers. Active routers are located in critical places of the network. For video distribution, these routers are to be accessed by registered video service providers, who download appropriate adaptation policy protocols. Protocols are written in the Programming Language for Active Networks (PLAN).[7] This approach provides both safety and efficiency. The server transmits the video at a constant bit rate to avoid burstiness. Within the network, active routers adapt the video stream on a per segment basis to the amount of traffic that a given segment can correctly deliver at the current time. Active routers also manage client subscriptions. An active router can thus further reduce network traffic by only sending a video stream to the parts of the network in which there are subscribed clients.

10.3.1 Adaptation Policy

The goal of adaptation policy is to reduce the bandwidth of the MPEG stream so as to avoid segment congestion. Adaptation is performed by frame dropping. Because MPEG relies on a coding of images into inter-dependent frames, randomly dropping one frame may prevent the decod-ing of several neighboring frames, thus seriously degrading the video quality. Thus, the frames to be dropped must be carefully selected. There are three kinds of frames in an MPEG stream: I, P, and B. Only an I frame contains a complete image and can be decoded independently. Decoding a P frame relies on the result of decoding the previous I or P frame. Decoding a B frame relies on the decoding of the nearest previous and successive I or P frame. An I frame and all of the frames that follow it, up to the next I frame, can be considered a single unit, known as a *group of pictures* (GOP). These frame interdependencies imply that the highest priority should be given to the transmission of I frames. Because a B frame does not influence the decoding of any other frame, transmission of B frames has the lowest priority. The strategy for dropping frames depends on the per second bandwidth requirements B_i, B_p, and B_b of I, P, and B frames, respectively, as well as the available bandwidth. Let us consider a frequently used GOP structure: IBBPBBPBBPBBPBB. According to this structure and the relative priority of the types of MPEG frames, we use the following strategies:

> **I frames**: The protocol never drops an I frame.
> **P frames**: Once a P frame has been dropped, the rest of the GOP cannot be decoded. Thus, P frames are only slightly less important than I frames. The protocol only drops a P frame if a previous P frame in the current GOP has been dropped or the available bandwidth is less than the bandwidth requirement of P frames.
> **B frames**: If the protocol has dropped a P frame in the current GOP, the B frame is dropped. Otherwise, based on the low priority of B frames, this frame is only transmitted if the available bandwidth allows both P and I frames to be transmitted as well. This scheme improves the quality of video transmission.

The active gateway[14,15] network architecture for videoconferencing consists of a set of active application gateways, a set of end systems, and a set of dispatchers, all of which can be placed on arbitrary network nodes.

Figure 10.2 shows a simple but typical active network setup where an active gateway is placed on a network node between two networks, each connecting an end system. Also connected to the gateway via a separate network is a dispatcher, which serves as an operational platform for the active gateway. The dispatcher issues programs to the gateway and receives messages that report application status, resource uses, statistics

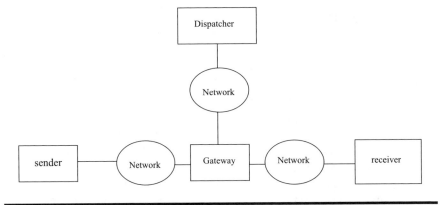

Figure 10.2   Active gateway network architecture for videoconferencing.

about network traffic, etc. The active gateway performs traditional packet routing and executes programs dispatched to it. Because dispatchers and active gateways are implemented at the application level, both can be launched as stand-alone applications from any node and can run on any target node in the network, allowing easy installation and utilization by end users. In practice, only one dispatcher is needed because it can concurrently dispatch various programs to as many gateways as possible.

10.4 The Active Traffic and Congestion Control Mechanisms

Internet applications and multimedia systems such as multimedia conferencing and distance learning are of great importance these days. These require large network bandwidth for real-time multicast and efficient adjustment of network operation for supporting heterogeneous receivers. This is because network conditions fluctuate significantly in terms of available bandwidth. The existing multicast protocols mostly consider the cost of an end-to-end path and support best-effort services. Therefore, they cannot accommodate dissimilar requirements of heterogeneous receivers. Various mechanisms exist for adjusting the transmission rate of senders in accordance with the network congestion condition. Due to the heterogeneity of the Internet, a single transmission rate from a sender cannot satisfy the conflicting bandwidth requirements at different sites. Consequently, the transmission rate is usually decided according to the receiver with the smallest bandwidth. This results in the degradation of the data received at other sites.

In Active Traffic Control Mechanism for Layered Multimedia Multicast (ATLM),[10] this is achieved by dynamically controlling the transmission rate at each router according to the network congestion status, which is provided by immediate children nodes. The mechanism is implemented in the active routers. The active routers in ATLM are similar to the core routers in a multi-core-based tree.

In ATLM, each active router or end receiver in the network monitors the loss rate of incoming packets on each link. If the packet loss rate exceeds a designated value, the routers notify their immediate active parent nodes (they can be active routers or senders) of the congestion condition. The active parent node then reduces the transmission rate of the link from which the notification message arrives. The change in the transmission rate at each link does not affect the condition of other links. It does not lead to any significant changes in the QoS at the receivers either. The ATLM scheme consists of two major parts: *traffic monitoring* and *adaptation*. Figure 10.3 shows the overall structure of the ATLM.

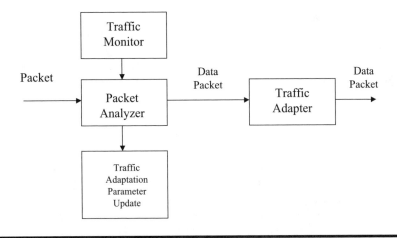

Figure 10.3   The structure of the ATLM.

10.4.1 Traffic Monitoring

When a packet arrives at an active node, the *packet analyzer* analyzes it. If it is a traffic state report packet, the active router updates *traffic adaptation parameters*. If it is a data packet, it is transferred to the *traffic adapter* and filtered according to the traffic condition of the downstream links, which is maintained in the *traffic control database*.[10] The active node also monitors the incoming traffic and decides the packet loss rate. Additionally, it sends a traffic state report message to the parent active router. If the packet loss rate exceeds the upper threshold, the active node regards the link as congested. Then the active node sends a message to its parent node in the multicast tree. The message indicates the congestion condition, and thus the parent node reduces the amount of traffic directed to the link from which the message arrives. In contrast, if the packet loss rate is lower than the lower threshold, then the active node sends a message notifying that it is in the unloaded state.

10.4.2 Traffic Adaptation

When an active router receives a traffic state message from its child active router, it updates its traffic adaptation parameters, which determine the transmission rate on each respective output link. If the traffic state message indicates congestion, the active router increases the filtering rate. In contrast, when the traffic state report message indicates an unloaded state, the active router increases the amount of traffic by decreasing the filtering rate. Because the network traffic changes continuously, the repetitive adjustment process goes on continuously.[10]

Active congestion control (ACC)[12] is an improvement to feedback congestion control that uses active networking techniques to shorten the control loop and improve network performance. The system improves feedback performance in a high-bandwidth delay product network with bursty traffic. ACC makes feedback congestion control a more distributed process by including router actions as well as endpoint actions. ACC modifies traditional feedback systems by adding active applications (AAs) that detect congestion, modify packet flows in the network to relieve congestion, and finally notify endpoints of modifications to their traffic. Unlike endpoint-only congestion control systems, the response begins at the congestion point and propagates to the endpoint, resulting in a quicker response time and shorter congestion periods.

Protocols such as the Internet Transmission Control Protocol (TCP) and Asynchronous Transfer Mode (ATM) cell-based protocols start to show degraded performance in the face of rapidly changing traffic characteristics in a network with a high-bandwidth delay product. This combination is difficult to control because the conditions change more quickly than the systems at the end can change their behavior to compensate.

When an ACC router becomes congested, it chooses one of the traffic streams passing through it and edits that stream's traffic so that it appears to the router as though the endpoint has responded immediately.[12] As a result, the correction appears at the router more quickly than the congested router could have communicated such information to the endpoint. The router edits the traffic by asking the upstream router to filter out packets from the chosen endpoint. A further notification of this event is sent to the endpoint, which adjusts its future behavior, for example, by reducing its sending window and marking the unsent packets to be negatively acknowledged. These congestion actions are taken by AAs inserted by ACC into the programmable elements of the network. An AA, in the Defense Advanced Research Projects Agency (DARPA) active networking architecture, is any small, secure program running in the network that has access to a variety of services from the router. ACC AAs require access to the queuing levels of the router and to the header information of packets to be discarded by congestion, and the ability to set filters to remove specific packets from the data stream. ACC is not a specific feedback system, but an extension to existing systems.[12]

10.5 Summary

The Internet has witnessed unprecedented success. Rapid growth in the number of users and the everlasting improvements in network infrastructure are leading toward a scenario where new and more powerful services are required. This chapter has discussed different applications using active and programmable networks to facilitate this idea.

Active e-mail provides context awareness-based support for customers. The network dynamically recognizes the customer's context. The user's context includes not only the location, but also capabilities of devices around the user that can be controlled. The network dynamically discovers the user's context at that specific moment.

Self-organizing video streaming dynamically adapts in real-time to the available bandwidth to ensure that each client receives the video at an acceptable level of quality.

The adaptation policy to transport MPEG2 over the Internet reduces the bandwidth of the MPEG stream so as to avoid segment congestion. Adaptation is achieved through the process of frame dropping.

The Active Traffic Control Mechanism for Layered Multimedia Multicast (ATLM) dynamically controls the transmission rate at each active router according to the network congestion conditions.

Active congestion control (ACC) improves traditional feedback systems by adding active applications (AAs) that detect congestion, modify packet flows in the network to relieve congestion, and finally notify endpoints of modifications to their traffic.

Exercises

1. Discuss the working of active electronic e-mail.
2. How do active routers enhance functionality to transport video over the Internet?
3. In what ways is the active congestion control mechanism different from traditional congestion control schemes?

References

1. D.J. Wetherall and D.L. Tennenhouse, The ACTIVE IP option, Proceedings of the 7th Workshop on ACM SIGOPS European Workshop: System Support for Worldwide Applications, 33–40, 1996.
2. H. Dandekar, A. Purtell, and S. Schwab, AMP: Experiences with Building an Exokernel-Based Platform for Active Networking, Proceedings of the DARPA Active Networks Conference and Exposition (DANCE'02), *IEEE*, 77–91, 2002.
3. D. Larrabeiti, M. Calderón, A. Azcorra, and M. Urueña, A Practical Approach to Network-Based Processing, Proceedings of the Fourth Annual International Workshop on Active Middleware Services (AMS'02), 2002.

4. G. Coulson, G. Blair, D. Hutchison, A. Joolia, K. Lee, J. Eyama, A. Gomes, and Y. Ye, NETKIT: a software component-based approach to programmable networking, *ACM SIGCOMM Comput. Commun. Rev.*, 33, 55–66, 2003.

5. B. Williamson and C. Farrell, Independent active program representation using ASN.1, *ACM SIGCOMM Comput. Commun. Rev.*, 29:2, 69–88, 1999.

6. S. Kamouskos and A. Vasilakos, Active electronic mail, in *SAC 2002*, Madrid, Spain, 2002, pp. 801–806.

7. D. He, G. Muller, and J.L. Lawall, Distributing MPEG Movies over the Internet Using Programmable Networks, paper presented at the Proceedings of the 22nd International Conference on Distributed Computing Systems (ICDCS'02), 2002, pp. 161–170.

8. J.I. Khan, S.S. Yang, D. Patel, O. Komogortsev, W. Oh, Z. Guo, Q. Gu, and P. Mail, Resource Adaptive Netcentric Systems on Active Network: A Self-Organizing Video Stream That Automorphs Itself While in Transit via a Quasi-Active Network, paper presented at the Proceedings of the DARPA Active Networks Conference and Exposition (DANCE'02), 2002.

9. D. Reininger, D. Raychaudhuri, and M. Ott, A dynamic quality of service framework for video in broadband networks, *IEEE Network*, 12:6, 22–34, 1998.

10. S. Kang, H.Y. Youn, Y. Lee, D. Lee, and M. Kim, The Active Traffic Control Mechanism for Layered Multimedia Multicast in Active Network, Proceedings of the Eighth International Symposium on Modeling, Analysis and Simulation of Computer and Telecommunication Systems, San Francisco, CA, USA, 2000, pp. 325–332.

11. V. Galtier, K. Mills, Y. Carlinet, S. Bush, and A. Kulkarni, Predicting and Controlling Resource Usage in a Heterogeneous Active Network, paper presented at the Proceedings of the Third Annual International Workshop on Active Middleware Services (AMS'01), 2002.

12. T. Faber, Experience with Active Congestion Control, DARPA Active Networks Conference and Exposition (DANCE'02), 2002.

13. eXtensible Markup Language and other families' technologies, http://www.w3.org/TR/.

14. S. Li and B. Bhargava, Active Gateway: A Facility for Video Conferencing Traffic Control, paper presented at the Proceedings of the COMPSAC'97: 21st International Computer Software and Applications Conference, 1997.

15. S. Subramanian, P. Wang, R. Durairaj, J. Rasimas, F. Travostino, T. Lavian, and D. Hoang, Practical Active Network Services within Content-Aware Gateways, paper presented at the Proceedings of the DARPA Active Networks Conference and Exposition (DANCE'02), 2002.

Index